Nature's State:

Less ~~20%~~

13.56

Purchased b~~y~~

NATURE'S STATE

NATURE'S STATE

IMAGINING
ALASKA
AS THE LAST
FRONTIER

SUSAN KOLLIN

THE UNIVERSITY OF NORTH CAROLINA PRESS

CHAPEL HILL & LONDON

Designed by April Leidig-Higgins
Set in Minion by Keystone Typesetting, Inc.
Manufactured in the United States of America

The paper in this book meets the guidelines for permanence and durability of the Committee on Production Guidelines for Book Longevity of the Council on Library Resources.

Library of Congress Cataloging-in-Publication Data
Kollin, Susan. Nature's state: imagining Alaska as the
last frontier / Susan Kollin.
p. cm.— (Cultural studies of the United States)
Includes bibliographical references and index.
ISBN 0-8078-2645-6 (alk. paper)
ISBN 0-8078-4974-x (pbk.: alk. paper)
1. American literature—Alaska—History and
criticism 2. Environmental protection—Alaska—
Historiography. 3. Natural history—Alaska—
Historiography. 4. Frontier and pioneer life in
literature. 5. Frontier and pioneer life—Alaska.
6. Alaska—Historiography. 7. Alaska—In literature.
8. Nature in literature. I. Title. II. Series.
PS283.A4K65 2001   810.9'32798—dc21   2001027414

05 04 03 02 01   5 4 3 2 1

To my parents

A man could be a lover and defender of the wilderness without ever in his lifetime leaving the boundaries of asphalt, powerlines, and right-angled surfaces. We need wilderness whether or not we ever set foot in it. We need a refuge even though we may never need to go there. I may never in my life get to Alaska, for example, but I am grateful that it's there. We need the possibility of escape as surely as we need hope; without it the life of the cities would drive all men into crime or drugs or psychoanalysis.
—Edward Abbey,
   Desert Solitaire: A Season in the Wilderness

. . . Sever the flesh from my bones.
Hang them above a fireplace,
Frame the mounted head
In arctic fur
Or exotic plumage
Such as is seen only in zoos or
Left captive in rapidly dwindling
Rainforests.
—Mary TallMountain, "Once the Striped Quagga,"
   in *Listen to the Night: Poems for the Animal Spirits
   of Mother Earth*

CONTENTS

Preface xiii

INTRODUCTION
Inventing the Last Frontier 1

CHAPTER ONE
The Wild, Wild North:
Nature Writing, National Ecologies, and Alaska 23

CHAPTER TWO
Border Fictions:
Frontier Adventure and the Literature
of U.S. Expansion in Canada 59

CHAPTER THREE
Domestic Ecologies and the Making of Wilderness:
White Women, Nature Writing, and Alaska 91

CHAPTER FOUR
Beyond the Whiteness of Wilderness:
Alaska Native Writers and Environmental Sovereignty 127

CONCLUSION
Toward an Environmental Cultural Studies 161

Notes 179

Bibliography 199

Index 215

ILLUSTRATIONS

Alaska (map) xvii

Effects of the *Exxon Valdez* oil spill (editorial cartoon) 3

The oil spill shown as threatening an American "masterpiece"
(editorial cartoon) 4

Postcard showing Alaska superimposed across the "Lower 48" 9

U.S. weather map 10

World Explorer Cruiseline advertisement 16

The 800-mile-long TransAlaska Pipeline 48

Sign on pipeline 49

Pipeline with graffiti 50

Tourists at the Alyeska Pipeline Visitor Center 51

Federal and Native Corporation lands (map) 147

PREFACE

As a child growing up in Southeast Alaska during the 1970s, I developed a strong fascination with the state's sublime landscapes in both their "natural" and technological forms. Much of this attraction has to do with my having lived at the foot of a mountain within walking distance of a glacier whose size, we were always told, exceeded that of Rhode Island. Meanwhile, my interest in the state's built environment probably can be credited to the era in which I was growing up. One of my strongest memories attending Auke Bay Elementary School involves watching educational movies that featured the soon-to-be completed TransAlaska pipeline, an 800-mile-long structure that would eventually alter the state's natural and economic histories in profound and unimagined ways. Produced by the Atlantic Richfield Company, these documentaries showed the progress of construction as workers and machines labored together in subzero temperatures to complete one of the nation's modern-day industrial wonders. A striking image featured in these films and one that is still employed by the oil companies today involves a shot of the pipeline set against a majestic wilderness backdrop. The foreground of these shots always includes wildlife, usually a caribou or grizzly bear crossing the landscape, seeming to coexist peacefully and contentedly with this strange edifice. We get to have it all, these pictures tell us. Wilderness and industrial development. Scenic landscape and a booming economy. They all go naturally together.

Nature's State accounts for the ways Alaska has been imagined as a "natural" entity and the ways this understanding has helped constitute aspects of American national identity. Drawing on a variety of literary and historical sources, I argue here that the scenic and natural wonder that is Alaska is a deeply cultural phenomenon. The fact that at one time in its history the region was regarded as "Seward's Folly"—a frozen wasteland thought to be of little national importance until the efforts of late-nineteenth-century naturalists and tourists helped rethink this understanding—indicates the degree to which ideas of nature shift and change throughout history, and the ways these ideas are always tied up with larger social forces.

A central tenet shaping the field of cultural studies holds that scholars should always be aware of their own complicity with the topics they examine. It is thus appropriate that I address the ways I have been shaped by the forces I analyze here, having moved to the state as a young child with the already firm belief in mind that Alaska was an exotic landscape promising adventure and intrigue. In a place where 85 percent of the state's budget comes from oil revenues and where tourist dollars provide much of the rest, it is difficult to spend any time in Alaska and avoid the reaches of oil and recreationalist money. Even as the political economy of nature often co-opts progressive thought and action in the region, there are nevertheless many groups and individuals across the state who are struggling to recast how Alaska has been positioned in the nation's spatial imagination. I find it important, then, to give thanks and acknowledgment to all those persevering folk who continue to devote themselves to envisioning a different future for Alaska.

During the time I have spent working on this project, I have enjoyed the emotional and intellectual support of many colleagues, friends, and family in Bozeman, Minneapolis, and Juneau. Numerous faculty at Montana State University offered encouragement during the completion of this project. I especially thank Lisa Aldred, Simon Dixon, Dan Flory, Greg Keeler, Judy Keeler, Colleen Mack-Canty, Dale Martin, Mary Murphy, and Beth Quinn. As chair of the English department, Sara Jayne Steen offered advice at important moments and always managed to come through with clever solutions and much needed resources. Jim McMillan, dean of the College of Letters and Sciences at Montana State University, provided additional funding for this project as well as a research leave during a crucial time in writing this book. A Scholarship and Creativity Grant from the Office of the Vice President for Research and a Research and Creativity Grant from the College of Letters and Sciences helped facilitate the project's development. Thanks also to Jan Zauha and the dedicated librarians at MSU's Renne Library who helped me track down obscure references and interlibrary loan material, and to Josef Verbanac, who answered numerous computer questions for me. For enabling my research and writing, my appreciation goes to Beth Cusick. Over the years, the many students in my courses on "Environmental Cultural Studies," "American Travel Writing," "Literature of the American West," and "Literature and the Environment" asked challenging questions that made me clarify my thinking; thanks to them for helping me make new connections and insights about nature, culture, and national identity.

At the University of Minnesota, I was fortunate to have the support and encouragement of Professors Maria Damon, Paula Rabinowitz, and Marty Roth. Each of them read early drafts of this project and pushed me to extend my ideas in new directions. Susanne Dietzel, Jim Glassman, Maureen Heacock, Frieda Knobloch, Capper Nichols, and Thitiya Phaobtong provided excellent conversation on cultural studies, feminist criticism, and environmental theory. At the University of Minnesota, a Harold Leonard Memorial Film Study Fellowship enabled me to research parts of chapter 3 and the conclusion. Financial assistance for research and writing also came in the form of a Graduate School Special Grant from the University of Minnesota's Graduate School and a Senior Fellowship from the University of Minnesota's English department.

My editors, Sian Hunter, Pam Upton, and Suzanne Comer Bell, deserve thanks for facilitating the timely production of this book. Their carefulness, attention, and patience significantly shaped the final product. Mary Elizabeth Braun offered support at a crucial time in the completion of this book; that we were unable to work together through the final version is unfortunate. Happy trails to you, Mary! William Chaloupka, Simon Dixon, Dan Flory, Melody Graulich, Catherine Kollin, Thomas Kollin Sr., Dale Martin, T. V. Reed, Ken Ross, and Eric Sandeen read the manuscript in its entirety; thanks go to them for helping me make a much stronger argument. I, of course, am solely responsible for any mistakes here. Thanks also go to Sue Hodson at the Huntington Library and Ron Inouye at the University of Alaska–Fairbanks for their encouragement and help in locating materials for this book. I also received funding from the Huntington Library and the Canadian Embassy which aided in completing the research for chapter 2. Sections of this book were presented at various national conferences, including those sponsored by the American Studies Association, the Association for Canadian Studies in the United States, the Modern Language Association, and the Western Literature Association. Thanks to the audience members and panel commentators who pushed me to make stronger arguments. Parts of the introduction and chapter 1 appeared in *American Literary History* 12 (2000): 1–2. A portion of chapter 2 appeared in the *Canadian Review of American*

Studies 26 (1996): 2. Thanks to the journal editors, Oxford University Press, and the University of Calgary Press for allowing me to reprint this material here. Thanks to I. Milo Shepard for permission to reprint materials from the Jack London Papers at The Huntington Library. For permission to reprint poems from *The Droning Shaman* and *Life Woven with Song,* thanks to Nora Dauenhauer. Robert Davis also graciously granted permission to reprint from *Soul Catcher.* Thanks to Kitty Costello, trustee and executor of Mary TallMountain's literary estate, for permission to reprint poems from *Listen to the Night: Poems for the Animal Spirits of Mother Earth* and *Light on the Tent Wall: A Bridging.* Permission to reprint from *Light on the Tent Wall* was also granted by the American Indian Studies Center, UCLA, copyright © 1990, Regents of the University of California.

In Juneau, I shared many nights of great discussion with Tina Pasteris and Art Petersen. Over the years, Julie Van Driel filled my house with books. My parents, Thomas Kollin Sr. and Catherine Kollin, as well as my siblings, Thomas Kollin Jr., Nancy Kollin, and Charles Kollin II, provided childcare and much-needed diversion, all of which ensured that I had the time and energy to complete the research and writing for this book. My daughter, Michaela, always knew which books she would rather have me read. Thanks to her for rescuing me from this work at crucial times. Dan Flory, dedicated co-parent, incisive critic, and philosopher/film scholar extraordinaire, has been with this project from its inception. His patience, discipline, and intellectual insights have ensured that this book was completed more or less on time.

Finally, my appreciation goes to those places that continue to inspire this project—the Mendenhall Glacier, Thunder Mountain, Tracy Arm, Perseverence Trail, Sandy Beach, Auke Rec, Eagle Beach, the East Glacier Trail, South Franklin Avenue, Egan Drive, the Back Loop, and, last but not least, the End of the Road.

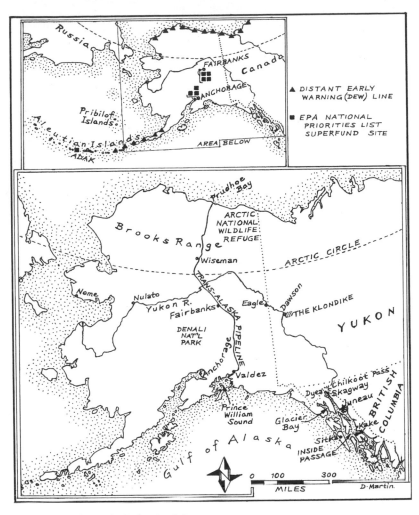

Alaska. (Map drawn by Dale Martin)

INTRODUCTION

INVENTING THE LAST FRONTIER

[O]ur powerful images of country and city have been ways
of responding to a whole social development. This is why, in
the end, we must not limit ourselves to their contrast but go
on to see their interrelations and through these the real
shape of the underlying crisis.
—Raymond Williams, *The Country and the City*

[L]andscape is an object of nostalgia in a postcolonial and
postmodern era, reflecting a time when metropolitan
cultures could imagine their destiny in an unbounded
"prospect" of endless appropriation and conquest.
—W. J. T. Mitchell, "Imperial Landscapes"

After the *Exxon Valdez* tanker spilled eleven million gallons of crude oil into Alaska's Prince William Sound during the spring of 1989, a great public outcry arose as the nation witnessed images of dying wildlife, oil-drenched beaches, and polluted seas on nightly television and front pages of newspapers. Considered one of the world's only remaining wilderness areas and one of its most popular tourist destinations, Alaska has been widely regarded as the "Last Frontier," a region whose history has yet to be written and whose "virgin lands" have yet to be explored. The oil spill threatened to disrupt Alaska's wilderness status, however, as Prince William Sound came to signify the profound environmental catastrophes facing the United States at the end of the twentieth century. According to many news reports, the *Exxon Valdez* disaster was most tragic because it took place in an area whose natural beauty was thought to surpass all others. If the region was a relatively unknown location for many Americans before the disaster, in the weeks and months after the spill, the media ensured that Prince William Sound became a household name through stories tracing its decline from a formerly pristine ecosystem into a place of extreme pollution.

Over the past two hundred years, however, intensive land and resource use in the form of mining, whaling, logging, and fox farming had altered this seemingly pristine region in rather dramatic ways. Even though forests in a secluded bay in Prince William Sound had been logged out by 1927 and second-growth timber had already been harvested by the time of the spill, the mainstream media nevertheless used images of this same bay to portray the sound as a once untrammeled but now endangered wilderness region.[1] While few reports questioned the accuracy of these depictions, a look at the oil industry's own record of operations in Alaska also indicates that the *Exxon Valdez* spill was just one of many environmental disasters taking place in the history of Prince William Sound's industrial development. The 1989 disaster, for instance, marked the four hundredth spill in the region since oil began to be transported from the North Slope in Prudhoe Bay to the port of Valdez in Prince William Sound. Even when production in the North Slope began to decrease in the late 1980s, six hundred Alaskan spills were being reported each year, prompting the EPA to name several locations in the area as severely contaminated.[2] In the aftermath of the *Exxon Valdez* disaster, however, most mainstream news reports ignored the history of the region's economic develop-

This editorial cartoon shows the effects of the *Exxon Valdez* oil spill on consumers of Alaskan seafood. (Reprinted by permission from Steve Kelly and Copley New Services; collection of the Alaska and Polar Regions Department, Rasmuson Library, University of Alaska, Fairbanks)

ment, and instead produced stories lamenting the destruction of one of the world's last remaining wilderness areas.

I highlight these points not because I wish to suggest that the oil spill was a small mishap that should be quickly brushed aside or that a strong public outcry was somehow unjustified; rather, I draw attention to these points in order to foreground the ways Alaska was shaped in the weeks, months, and years after the disaster. When we consider the rhetoric of the outcry following the spill, it becomes apparent that something else was at stake in discussions of the Alaskan landscape for, in addition to destroying populations of wildlife and polluting vast areas of sea and land, the disaster also threatened the meanings and values assigned to Alaska in the popular national imagination, understandings that were not necessarily shared by indigenous populations across the state. Patrick Daley has examined the media coverage of the spill, comparing the ways Alaska Native and nonnative newspapers defined the event. He argues that for the most part, main-

Another editorial cartoon depicts the oil spill as threatening an American "masterpiece." (Reprinted by permission from Mike Luckovich and Creators Syndicate; collection of the Alaska and Polar Regions Department, Rasmuson Library, University of Alaska, Fairbanks)

stream reports relied on conventions of the disaster narrative, seeking to lessen confusion and assure the public that the danger was contained. Later, these news reports began concentrating on the human-interest aspect, voicing the frustrations of local residents who were disillusioned with the oil industry, and with state and federal governments. The people who predominantly figured in this coverage were white commercial fishermen from Valdez and Cordova, cities located between twenty-eight and fifty miles from the spill.

With this focus, however, the mainstream press ended up neglecting the disaster's impact on Prince William Sound's largest private landowners—the Alaska Natives. Particularly absent were the voices of the Aleut villagers of Tatilek who resided only six miles from the spill's origin. In contrast to mainstream reports that featured images of oil-drenched birds and damaged beaches, symbols of a once untrammeled nature now polluted, Alaska Native reports focused on other, less sentimental, concerns. *Tundra Times*, a weekly newspaper dedicated to the concerns of the state's indigenous population, primarily addressed the impact of the spill on subsistence issues; instead of featuring dramatic images of oil-soaked wildlife, *Tundra Times* told of villagers' attempts to cope with the crisis now facing a population for whom nearly 50 percent of its food is harvested from the sea and the land. I point to this study and its comparison of news coverage of the disaster for the ways it highlights competing responses to the oil spill and for the ways it foregrounds contested understandings of human relations with nonhuman nature.[3]

These different understandings may be noted in the way Alaska has been situated as a sublime wilderness area in the nation's spatial imagination.[4] Widely regarded as the Last Frontier, Alaska is positioned to encode the nation's future, serving to reopen the western American frontier that Frederick Jackson Turner declared closed in the 1890s. In this sense Alaska functions as a national salvation whose existence alleviates fears about the inevitable environmental doom the United States seems to face and, like previous American frontiers, promises to provide the nation with opportunities for renewal.[5] This notion of regeneration and renewal is not merely psychological but also involves an economic dimension. With its vast timber, fishing, oil, and mineral reserves, Alaska is thought of as a land endowed with great natural wealth, a terrain offering unlimited commercial opportunities. As a result, it is considered one of the few remaining areas where the United States may enact what Richard Slotkin calls "bonanza economics," the acquisition of abundant natural resources without equal inputs of labor and investment.[6]

During a time when discussions about the environment are repeatedly framed by what Alexander Wilson calls a discourse of "crisis and catastrophe," caring for "wilderness" areas such as Alaska becomes an increasingly urgent project.[7] I would contend, however, that this anxiety-laden rhetoric does not merely reflect concerns about nature but, more important, signals concerns about U.S. national identity.[8] The discursive construction of Alaska as the Last Frontier marks a yearning for undeveloped lands in a world whose surfaces are perceived to be fully mapped.[9] Because the urge to protect a Last Frontier points both to anxieties about the environment and to concerns about the nation's status and future, Alaska emerges as an object whose production is linked to the United States' identity and expansionist history.

Nature's State enters current debates and discussions about the U.S. environmental imagination, tracing the ways Alaska, a seemingly liminal space in American culture, functions to alleviate larger anxieties about nature and national identity. While Alaska often fascinates Americans because of its status as the Last Frontier, the region nevertheless remains largely outside the United States' imagined community, serving as an extraneous space not fully accommodated into a national sense of self. Like many U.S. weather maps that feature a greatly reduced version of Alaska next to southern California, most studies of American literature, history, and culture also consign the state to their margins, if they include it at all. *Nature's State* addresses this lack and examines the ways popular responses toward Alaska foreground larger concerns about the meaning of America while also illuminating important aspects about the connections between nature advocacy, landscape conventions, and U.S. national development.

Here I argue that the environmental discourses shaping Alaska cannot be separated from the nation's larger expansionist concerns and its history of development. As much as it might be tempting to consider the protection of nature and the expansion of nations as radically different projects, they have frequently worked hand in hand in defining specific uses in which to employ Alaska. The Euro-American drive for continental control during the late nineteenth century, for instance, situated the Far North as a strategic frontier for national expansion. Because of its position in the Pacific near Russia, Canada, and the Arctic, Alaska emerged as an important geopolitical site for national security and foreign policy concerns. At the turn of the century, however, a movement also arose that expressed concerns about the proper uses of the national landscape. During this period, frontier discourses helped shape environmentalist rhetoric, and ecological projects became closely linked to U.S. expansionist enterprises.

Popular perceptions of Alaska as the Last Frontier stem from two nineteenth-century developments: the United States' overseas expansion and the Progressive-era conservation movement. During a time when it was widely believed that the internal American frontier faced exhaustion, both projects had great stakes in locating a new frontier. Although historians have indicated that the "closing of the frontier" was actually a myth, the idea nevertheless held great power in the national imagination, providing yet one more rationale for the United States' expansionist projects.[10] After the Civil War, the nation began creating an overseas empire designed to curb the problems of agrarian and industrial overproduction and the closing of the frontier. As the primary architect of what Walter LaFeber calls the "New Empire,"

Secretary of State William Henry Seward negotiated the purchase of Alaska from Russia in 1867 in order to secure a strategic base for the nation's overseas markets.[11] Alaska was intended to be the first in a long line of territorial acquisitions the United States sought in creating a "security perimeter" of American-owned lands in the Pacific.[12]

Alaska's position on the nation's map, however, also threatens to dismantle myths of national identity; in particular, it calls into question notions of American exceptionalism, as the region's location tends to highlight the entangling alliances of imperialism the nation engages in but continues to repudiate. Historian Morgan Sherwood, for instance, once wrote that Alaska's separation from the continental United States made it a "freak in American development—the first important colonialist effort of the European variety."[13] While we could argue that each attempt to include new lands, continental or otherwise, in fact signals an instance of U.S. imperialism, Sherwood's remarks concerning Alaska's geographical position are nevertheless quite telling.[14] In U.S. national narratives, it is typically assumed that American development is continual rather than disconnected; Alaska's position on the map thus appears "freakish" to Sherwood and others because the region cannot be narrated through the accepted paradigms of American national expansion. The geographical continuity that Sherwood seeks mystifies the fact that each incorporation of new lands marks yet another moment in the history of U.S. nation building. The discontinuity between Alaska and the rest of the United States appears "freakish" to him, in part, because it signifies a rupture that the nation continually attempts to ignore. During the purchase of Alaska, for instance, Secretary of State William Henry Seward and other nation builders sought to annex British Columbia along with Alaska in an effort to secure continuous U.S. rule throughout the hemisphere. Alaska was envisioned as the first in a long line of territorial conquests that would include Canada and Latin America.[15] As the gap on the charts indicates, however, efforts to expand the nation's borders up through Canada were defeated. The separation between Alaska and the rest of the country thus points to a failed moment in this particular expansionist project and signals a disruption in the nation's efforts to gain continental control.[16]

Some national maps encode yet another response to Alaska. On certain U.S. charts, for instance, the region is depicted as a landscape whose tremendous size gives it great national importance. In an effort to highlight the significance of this territorial acquisition, maps such as those appearing in school textbooks or on postcards of the forty-ninth state often superimpose Alaska upon the continental United States, with the region's coastline stretching from the Pacific to the Atlantic Oceans. Because maps, as cultural geographers tell us, are never mere reflections of the world but operate as cultural documents that serve particular nationalist functions, charts that place Alaska across the continental United States are especially intriguing for the "patriotic allegory" they depict, revealing the nation's territorial imperatives and the value the United States places on expanding its geographical borders.[17] In this case, expansion involves incorporating a territory that comprises more than 586,000 square miles, roughly one-fifth the area of the rest of the United States. By highlighting the region as an important source of national pride, these maps more fully serve the needs of American national development, foregrounding the United States' ability to extend its borders and renew itself once again.

Close readings of the nation's maps indicate, however, that while some charts celebrate Alaska as an important national entity, other maps present the state as an almost insignificant terrain. Alaska's national importance, for instance, diminishes on many U.S. maps that present a greatly reduced version of the state next to southern California. Hawai'i receives a similar fate as it, too, is frequently transplanted from its geographical setting in order to be placed in closer proximity to the rest of the United States. These efforts involve more than merely saving space on the chart; instead, the diminished distances between Alaska, Hawai'i, and the rest of the United States serve larger purposes, for on these maps, the United States remains intact as one geographical entity which gives the nation an appearance of territorial stability. Alaska's shifting position on the nation's maps thus presents us with multiple versions of a national epic. While one map indicates a desire for empire by proudly displaying the fruits of the United States' territorial conquests, the other map demonstrates a denial of empire by reducing the evidence of the nation's imperialist past. Although the nation's westward expansion from the Atlantic to the Pacific may be regarded as the appropriate course of empire, the acquisition of Alaska and other lands not geographically connected to the continental United States threatens to expose the nation's imperialist gestures, for on these charts, the gap signifies empire in a way that the continuous line from coast to coast doesn't.

Sherwood once described the United States' expansionist interests in the Far North; according to him, Alaska served as a "bridge be-

A postcard features Alaska superimposed across the Lower 48. (Reprinted by permission from Arctic Circle Enterprises)

tween the Old World and the New," "a fertile, strategic, and virgin ground," and "an unused laboratory that promised to yield evidence confirming or refuting some of the [world's] great theories."[18] Indeed, Alaska's geographical position has inspired many ambitious transnational projects. One such enterprise conducted in 1867 by the Western Union Telegraph Company sought to build a submarine cableline from Alaska, over the Bering Straits, and across Russia in an effort to create an international communication system to facilitate U.S. interests in Asia. The project was abandoned, however, with the installation of the Atlantic line. Another ambitious scheme was developed in 1899 by the railroad magnate Edward H. Harriman, whose dreams of constructing a New York–to–Paris route involved an excursion through the same region.[19] In both of these instances, the geopolitical position of Alaska provided the means through which the New World could map itself onto the Old World.

Just as the legacy of U.S. expansion helped position Alaska as a Last Frontier, the Progressive-era conservation movement also played a role in shaping ideas about the region. With the perceived closing of the frontier and the realization that the nation's natural resources could not be endlessly extracted, conservationists sought ways of curtailing uncontrolled development and establishing more efficient uses of natural resources.[20] The Progressive-era conservation movement

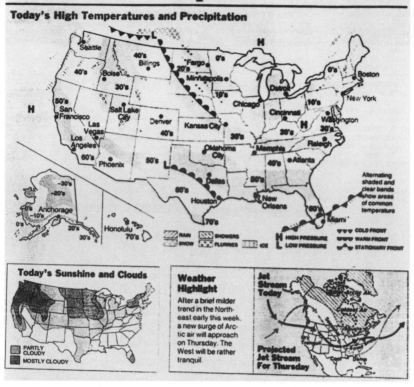

U.S. weather map featuring a much-reduced Alaska relocated near southern California. (Reprinted by permission from the *New York Times*)

eventually splintered, with advocates of conservationism favoring carefully planned resource development and supporters of preservationism arguing that the nation's wild landscapes should be protected from such uses.[21] Although they differed from each other in certain ways, the two groups nevertheless shared much in common: arising out of what appeared to be the closing of the frontier, both of them looked to the environment as a way of solving the crisis of exhaustion facing the United States. The two movements also converged in other ways: while conservationists valued nature for its economic uses and preservationists valued nature for its aesthetic and amenity uses, both groups considered nature a valuable commodity in an era when U.S. landscapes were increasingly at risk.[22] Recently scholars have begun addressing this other history of U.S. environmentalism, restoring to memory the underside of nineteenth- and twentieth-century concerns for nature, and relating them to larger projects of expansion. Environmental historians, for instance, have chronicled the removal of American Indians from their homelands in the establishment of wilderness regions and national parks across the nation, and from there have gone on to describe new forms of environmental interventions in other parts of the world that strongly resemble older forms of colonialism.[23] Efforts to protect, conserve, and wisely use nature— whether it be "our" nature or "their" nature—have been central elements in the production of American identity, and for that reason deserve to be reexamined for the larger concerns they address.

Beauty Marks and Oil Spills

Today, constructions of Alaska as the Last Frontier continue to perform a crucial function in the national imaginary. Melody Webb, for example, investigates recent depictions of the Last Frontier, pointing out that these portrayals of Alaska emerged in the 1970s as a state slogan and advertising ploy used to romanticize the nation's expansionist history. Webb contends that images of Alaska as a Last Frontier helped resituate the frontier past in the present era, nostalgically enlisting the region as a new site for national myths about the winning of the West.[24] Depictions of Alaska as a mythic frontier space also encode other meanings in the Euro-American imagination, appearing as a rallying cry for environmentalists concerned about threats to the nation's landscapes. William Goetzmann and Kay Sloan argue, for instance, that competing impulses shape the Far North which result in what they call the "two Alaskas," a vision of the region as "a wilderness to be preserved and a frontier to be exploited."[25] Rather than viewing these responses as antithetical to one another, however, we might instead consider the ways they operate as mutually sustaining ideas. Mainstream environmentalist rhetoric often advances notions of American exceptionalism, masking the nation's expansionist desires in myths of the United States as a benevolent international force, the protector of imperiled landscapes and populations alike. These desires continue to shape various environmentalist efforts, surfacing in recent efforts to protect wild regions such as the Brazilian rainforest, or in responses to acts of ecological destruction such as the

Kuwaiti oil wells set on fire during the Persian Gulf War.[26] In these instances, U.S. concerns about the environment, especially other peoples' environments, are typically presented as an ecofriendly gesture.

Environmentalist rhetoric of this sort frequently obscures development ideologies in other ways as well. During the late nineteenth and early twentieth centuries, when industrial development had extended across the nation, environmentalists responded by demanding the protection of natural landscapes for nonindustrial uses. Alexander Wilson points out that in hindsight these nonindustrial uses have largely turned out to be tourism, an activity that shares much in common with the imperialist adventure by advocating an "unquenchable appetite for the exotic and the 'uncharted.' "[27] Wilderness protection itself has often contributed to a lucrative tourism business, transforming landscapes in ways that can be as destructive to the environment as other forms of industrial development.[28] During the spring of 1989, for instance, the *Exxon Valdez* disaster challenged the nation's image as a protector of the environment and threatened to disrupt Alaska's status as one of the most desired destinations for international tourism. The crisis management following the disaster therefore faced the double duty of cleaning up both the spill and Alaska's soiled image. While the country struggled to make sense of the disaster, the Alaska tourism industry acted quickly, devising a public relations campaign intended to counter the spill's negative effects. One full-page text appearing in several major U.S. newspapers featured an image of Hollywood movie star Marilyn Monroe. "We've changed this picture to make a point about a legendary beauty," the line across the top of the page reads. In smaller print, readers are told that Monroe's beauty mark has been removed from her cheek, a process that is then likened to the aftermath of the oil spill. The text explains:

> Unless you look long and hard, you probably won't notice her beauty mark is missing. Without it, the picture may have changed, but her beauty hasn't. The same is true of Alaska. The oil spill may have temporarily changed a small part of the picture, but the things you come here to see and do are as beautiful as ever.

Designed to minimize the damaging effects of the oil spill, the ad presents the spill as a blemish, as if it were merely a small aspect of a larger picture, a calculated act instead of an environmental disaster. Through this analogy, the tourist industry portrays the oil spill as changing only a "small part of the picture," causing only "temporary changes" to the wildlife and environment which are supposedly barely noticeable to visitors. The lack depicted here in the form of Marilyn Monroe's missing beauty mark is regarded as inconsequential, a loss that will hardly be registered by viewers. The ad, however, reverses what had actually taken place after the disaster: in a more logical analogy, the beauty mark should become larger, encompassing a greater area of Monroe's face, rather than simply disappearing with the aid of an artist's airbrush.

As Donna Haraway has suggested, much is at stake in our judgments about nature, in what counts as nature or culture, and who gets to inhabit natural categories. Arguments from nature are often absolutely central to larger societal debates, especially those concerned with race, gender, and class.[29] In this way, the ad taps into many contemporary cultural discourses, engaging its meaning by employing not just any Hollywood film star, but Marilyn Monroe, the nostalgic icon of white femininity and glamour during the 1950s. This use of an unambiguously white actress as the embodiment of the Alaskan landscape ultimately effaces the state's native populations by naturalizing whiteness as the proper sign of the racialized landscape.[30] The advertisement also advances problematic notions of gender by presenting Monroe as an enticing come-on, a figure beckoning readers to enjoy what she has to offer. According to the text, the movie star's attraction lies in her seemingly abundant sexuality, a quality apparently mirrored in Alaska, whose own lush landscapes indicate that the region is safe from ever being completely destroyed. In other words, the endless abundance associated with both entities suggests that there's enough of Alaska and Marilyn Monroe for everyone to enjoy.

The decision to enlist the actress in efforts to reglamourize Alaska may seem an odd choice for an ad campaign charged with the task of promoting a landscape in need of renewal. After all, not only has Marilyn Monroe's own image been threatened by scandals over the years, but her natural beauty was also highly constructed. Yet, in spite of or perhaps because of it all, Monroe continues to be regarded as desirable, a legend who survives in American minds despite her death over three decades ago. Obviously the industry hopes the same will be true of Alaska, for according to the ad, the cleanup efforts following the spill enable visitors to experience what they have always enjoyed about the region. "From the Kenai Peninsula to Ketchikan, Mt. McKinley to Columbia Glacier in Prince William Sound, Alaska is still home to spectacular fishing, plentiful wildlife and untold adventure. So come to Alaska this summer. And see one of the most beautiful faces on earth."

What is likewise intriguing about the ad are the ways it foregrounds how U.S. expansionist concerns continue to shape the region in the dominant national imagination. By suggesting that Alaska's beauty has been restored during cleanup, the advertisement displays aspects of what John McClure calls "redemptive unmapping," a response arising in the era of late imperialism when the West began to perceive that new global conquests were no longer available to be claimed. Redemptive unmapping enables the West to empty landscapes of their signs, not in an effort to leave them untouched, but in order to recreate the conditions for their own remappings. The tourist ad engages this process by erasing Marilyn Monroe's beauty mark and the effects of the *Exxon Valdez* oil spill. In describing these marks as coming off rather than going on, the ad seeks to return the landscape to a pre-discursive moment, to a time when "untold adventures" and other acts of exploration may still be experienced. The ad campaign thus contributes to a project of unmapping by allowing Euro-Americans the opportunity to wipe the slate clean so that they can secure the means for their own inscriptions in the present era.[31]

Because much of the post-disaster news coverage engaged in a project of remapping that obscured the area's history of development, few people noted that the outcry following the spill in some ways mirrored responses that surfaced during the eighteenth century when European explorers sailed to the area and found that, even then, the region was not the unspoiled terrain they sought. In 1778, for instance, when Captain James Cook arrived off the coast of present-day Alaska at a site he later named "Prince William's Sound," the British explorer remarked that the land already showed signs of European contact. During his stay in the area, Cook noted that the natives he encountered appeared familiar with western goods even though to his knowledge they had never had direct contact with Europeans. According to Cook, the natives indicated their desire for trade but also explained that they already had a surplus of certain European goods such as copper, by "pointing to their weapons, as if to say, that having so much of this metal for their own, they wanted no more."[32] In his travel account, Cook thus expressed his dismay that the region and the people he encountered had already been altered by the political economy of European expansion before any of the natives had actually met its representatives in the flesh.

A decade later, the British explorer George Vancouver returned to Prince William Sound, noting that great changes had taken place since his previous voyage with Cook. Vancouver noted, for instance, that during his surveying excursions, "not a single sea otter and but very few whales and seals had been seen." He also remarked that "Many trees had been cut down since these regions had been first visited by Europeans," a sign indicated "by the visible effects of the axe and saw."[33] As information about European voyages in Prince William Sound became public during the latter part of the eighteenth century, other explorers quickly arrived in the region, hoping to reap benefits from the area's natural resources. Traffic off the coast of Alaska soon reached such heights that one English commander who sailed to nearby Cook Inlet in 1786 responded to the sight of other explorers with dismay, noting that "some nation or other had got to this place before us, which mortified me not a little!"[34] Another British explorer who arrived in Prince William Sound the following summer also bitterly remarked that the presence of other European merchants in the region served as a "*coup de grace* to my future Prospect of Success in this line of Life."[35]

Like their eighteenth-century predecessors, today's travelers continue to lament their belated arrival in the North, and in the spring of 1992, one post-spill tourist ad tried to capitalize on this anxiety. Appearing in popular magazines three years after the *Exxon Valdez* disaster, a World Explorer Cruiseline advertisement described Alaska as the "final American frontier," a place that is "meant to be explored." The cruiseline promised to provide tourists who sail on the "S.S. Universe" a "full . . . 14-day adventure for the heart, mind and soul." Adopting the rhetoric of discovery and exploration, the cruise ship ad captures once again how depictions of Alaska often rely on frontier nostalgia and anxieties of belatedness, advancing ideas about the region as a site for adventure and intrigue in the modern era.[36]

Although the spill marked an important rupture in the nation's ideas about Alaska, five years after the spill, tourism industry figures showed that travel to Alaska actually increased dramatically in the years following the disaster, with the number of tourists visiting the state on cruise ships alone rising more than 50 percent.[37] Rather than curtailing interest in Alaska, the 1989 disaster may have had the op-

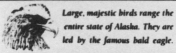

World Explorer Cruises advertisement featuring trips to the "final American frontier." (Reprinted by permission from World Explorer Cruises)

posite effect, as news coverage of the spill made travelers more anxious to visit the Last Frontier. During the first months following the disaster, some tourist agencies even advised travelers to book ahead for passages to Alaska in the following summer.[38] The desire to visit the state thus remained powerful even after the *Exxon Valdez* disaster complicated the region's status as a wilderness area. The spill, in fact, may have accelerated tourists' sense of urgency to visit Alaska as the environmental catastrophe itself became incorporated into the spectacle of the Last Frontier.[39]

Green Ideology

Although Alaska has been valued, appreciated, and marketed to potential consumers as a "state of nature," this study proceeds with the assumption that the region should be understood less as a natural terrain than a socially constituted entity. In this sense, Alaska's status as a natural wilderness area might be best understood as a sign of its intense and even unrelenting cultural production. In an era facing a rapid depletion of natural resources, places regarded as wilderness areas typically serve important national functions. Wilderness regions are often regarded as sanctuaries, as sites that allow us to escape from the realm of human activity. The concept of wilderness, however, needs to be reassessed, for, as several critics tell us, the very project of assigning certain areas wilderness status inevitably requires that they become incorporated into the domain of culture, an act that transforms wild areas into products of human design.[40] We see this process take place today as wilderness itself has become a strong growth industry, a valuable commodity that may be packaged, marketed, and sold in a postmodern era nostalgic for pristine or unspoiled lands.[41] Far from being places of escape from the social world, wilderness areas are also governed by unwritten but very real rules that dictate the types of behavior acceptable in those spaces. In that sense, the social protocols associated with wilderness regions are often as detailed and elaborate as those associated with urban spaces.

During the aftermath of the disaster, the region and the nation's environmental health appeared grim, and the spill seemed most disturbing because it occurred in one of the nation's most valuable wilderness areas. In an age of ecology, natural sites are assumed to have greater value than places that are lived in, and certain areas themselves become invented and reinvented as wilderness sites. Such was the case with Alaska during the wake of the disaster. The project of restoring the landscape to a pristine state (as if that were even possible) resulted in various misguided efforts, one of which included the use of highly toxic chemicals to break down oil deposits in the ocean. Many biologists have since argued that these cleanup attempts actually had the opposite effect on the region's ecosystem, as some of the chemicals used proved to be even more damaging to marine life than the oil itself.[42] Efforts to clean up the spill also contributed to the destruction of the ecosystem in other ways by creating tremendous amounts of waste materials, including what one journalist called "ten tons of toxic garbage," much of which was burned in incinerators that polluted nearby Valdez.[43] It was estimated, too, that the spill ultimately created a rise in the GNP as more than $2 billion was spent on cleanup and another $40 billion on other damages incurred as a result of air pollution, amounts that would probably not have been reached had the disaster taken place in a less scenic part of the country.[44]

Yet if nothing else, the spill helped catapult environmentalism to a national public awareness with a force greater perhaps than even twenty years of Earth Day celebrations. The nation's response to the disaster, however, reveals the importance of examining how nature advocacy is constituted and absorbed into national culture. Timothy W. Luke argues that the popularity of environmentalism may not be the victory many would like it to be, but suggests instead that the recent "greening" of America might be better understood as a project orchestrated in large part by multinational corporations. After confronting a fierce battle waged by environmentalists during the 1970s, many companies began to cash in on the popularity of the green movement, mounting their own environmental strategies aimed at domesticating ecological radicalism. By the late 1980s, a new rhetoric of reform emerged which suggested that major corporations were no longer primarily responsible for pollution; instead, each individual consumer was now a major factor in determining the environmental health of the planet. As a result, what once had been radical environmental critiques were now reduced to a corporate ideology of green consumerism. The shift away from corporations has led to the emergence of what Luke calls an "ecological subject," a green consumer whose everyday economic activities now somehow play a crucial role in deciding Earth's fate.[45] Luke specifically points to the aftermath of the 1989 *Exxon Valdez* oil spill as an important moment in the production of this ecological subject.[46] Although the disaster had a great national impact, playing a central role in bringing environmental awareness to public attention, the event marked another turning point as Exxon's insidious and successful move to appropriate environmental discourses also made concerns about nature appear to be a politically safe project. For any study interested in examining the contours of U.S. environmental struggles, then, it becomes important to investigate the institutions and groups involved in advocating nature as well as the cultural work that is performed in the process.

A word about language is in order here. Because environmental awareness has long been incorporated into a national sense of self, it becomes important to examine what Langdon Winner calls "the text of nature," the ways social concerns shape nature as an object of interest in its own right.[47] Andrew Ross argues a similar point, calling attention to the ways the concept of nature operates as the "ultimate people-pleaser"; the sheer ease with which concerns about nature are disseminated in culture, often by the very forces associated with environmental destruction, should warn us about how nature can be made to speak for almost any political project.[48] Fears about the destruction of nature that reached a peak after the *Exxon Valdez* disaster must therefore be examined as part of a larger set of beliefs structuring national culture and national terrain. The environmental anxieties about Alaska following the spill highlight how frontier nostalgia and the yearning for natural plentitude continue to shape responses to North American landscapes in the present era.

Also, because landscapes are not naturally given, but rather are socially constituted entities whose meanings shift as the result of specific social practices, concepts such as the Last Frontier must be investigated for the ideologies they encode and the cultural work they perform. David W. Noble suggests that the frontier serves an important function in European American constructions of national identity. Envisioned as the "threshold . . . between a decadent Europe and a vital America," the concept enabled the ascendance of "American space" with its possibilities of renewal over "European time" with its inevitable forces of corruption. According to this national narrative, the North American landscape seemed to offer the United States limitless opportunities for expansion which supposedly enabled it to avoid the economic and social exhaustion other European countries faced.[49] The concept of the frontier also helped inaugurate the idea of American exceptionalism, the belief that the United States stands apart from the rest of the world as a unique national entity.[50] Amy Kaplan demonstrates that notions of American exceptionalism perform important ideological work, most notably by obscuring the study of U.S. imperialism. In depicting the United States as a unique national development, American exceptionalism forecloses the idea that the nation also engages in colonizing projects. Imperialism is thus projected onto other countries, Kaplan explains, as "something only they do and we do not."[51] In American cultural discourse, "frontier" often appears in the place of "empire," serving as a euphemism for the nation's acquisitionist activities. Europe may have empire, the story goes, but the United States has frontier, a term intended to signify a more politically innocent project. The frontier therefore functions as a complex rhetorical strategy that simultaneously advances the United States' desires for and denial of empire.[52]

During a time when the nation's projects of westward expansion and course of empire are being examined for their ideologies of conquest, frontier discourses are also undergoing reconsideration. In a discussion of Alaska's frontier status, Eric Heyne likewise suggests that we regard the frontier as a "rich textual" concept, a discursive space that is constantly reinvented as an aspect of the national obsession with exploration and conquest.[53] Michel Foucault likewise points out that many of our spatial metaphors including "region," "province," and "field" operate within a political realm. He reminds us, for instance, that "region" derives from "regere" which means to command, that "province" derives from "vincere" which refers to a conquered territory, and that "field" evokes the battlefield itself. One could also add "frontier" to this list, noting the way it refers to the front line of battle. For Foucault, the pervasiveness of these landscape tropes is a reminder of how the military has managed to "inscribe itself on the material soil and within forms of discourse."[54]

Because the concept of "wilderness" is also closely associated with Alaska in the dominant American imagination yet remains a highly contested word among scholars and activists, I think it is important to analyze the multiple meanings and ideologies embedded in this term. Throughout this study, I argue that the concept of wilderness, so central to the mainstream environmental movement, expresses assumptions about the national landscape that are exclusive to a Euro-American point of view. As environmental scholars contend, notions of a pristine or untouched natural landscape devoid of any human contact operate as an asocial concept that can only emerge by ignoring the history of native peoples in North America.[55] Many historians, in fact, now recognize that wilderness regions might be better understood as rural spaces, sparcely inhabited lands that are nevertheless occupied by human populations.[56] In this study, then, I use the term wilderness to speak of a socially constituted space that says more about the making of a Euro-American self than it does about any actual geographical landscape.[57]

Nature's State raises a series of questions about the nation's environmental concerns for Alaska. What is the dominant history of our thinking about nature, for instance, and how did this understanding affect the ways Americans responded to an event such as the *Exxon Valdez* oil spill? How did Alaska emerge as a pristine site that somehow has only recently been contaminated by human activity, and what purpose does this construction serve in the national imaginary? And finally, in what ways do popular understandings of Alaska as a wilderness area or Last Frontier actually share a great deal with other national narratives about the environment, especially with myths about the limitless abundance of the North American landscape? To answer these questions, I examine how ideas about nature connect to and help shape national identity, arguing that environmental anxieties have dominated thinking about Alaska from the time the United States purchased the territory from Russia in 1867. Many environmental historians point out that the ecological demise stemming from western expansion gave rise to an American environmental movement, with debates over resource development and water rights forming some of the movement's first concerns.[58] While this argument may lead us to adopt a narrowly defined understanding of environmentalism that overlooks struggles taking place in urban places located outside the West or that disregards the efforts of American Indians and other communities of color whose ecologies have not been central to the national effort, it is nevertheless important to investigate how certain social developments in the American West nevertheless enabled a *particular* type of environmental thinking to emerge.

Because the conquest of Alaska through discourses of expansion and environmentalism operates as an important chapter in U.S. history, debates about the Alaskan landscape may be investigated for the ways they serve larger national enterprises. In this way, the cultural production and reception of nature in the nation's spatial imagination becomes a crucial issue to examine in understanding Alaska. As Raymond Williams once explained, the idea of nature "always contains an extraordinary amount of human history."[59] While it often goes unnoticed, nature is perhaps culture's best invention. In the spirit of

Williams, then, *Nature's State* examines how ideas about the environment shape texts about Alaska, and at the same time help constitute Alaska as a text in its own right.

The title of this book recalls the work of Perry Miller, who half a century ago influenced the direction and development of American Studies, writing one of the field's landmark books, *Nature's Nation*. In his study, Miller described Nature as the "official faith of the United States," a major premise for white national self-consciousness.[60] The Euro-American identification of Nature with virtue, he argued, linked the very health and personality of the country to the fate of its landscape, and in doing so, set in play a whole series of problems that still plague national consciousness, from beliefs in an American exceptional identity to the persistant bias against urban spaces and urban culture in national life.[61] Although this "misguided cult of Nature" may have little real impact on how Euro-Americans actually behave toward the natural world, Miller pointed out that it nevertheless has had everything to do with how the populace *understands* its own conduct in the world. Miller thus warned about the need to assess what he regarded as the "sinister" dynamics operating behind national conceptions of Nature.[62]

Miller's observations are important to keep in mind in assessing depictions of Alaska as the Last Frontier. My aim in this book, then, is to address how the thinking that fetishizes Alaska as a pristine natural space connects to larger national preoccupations, making the region an important northern extension of "nature's nation." As scholars in environmental cultural studies contend, nature is always a highly contested entity that never merely speaks for itself. Geographer David Harvey argues along these lines, focusing in particular on the ways nature has always been central to processes of nation formation; as he points out, "nationalism without some appeal to environmental imagery and identity is a most unlikely configuration."[63] From the United States' purchase of Alaska from Russia in 1867 to the *Exxon Valdez* oil spill in 1989 and beyond, ideas of nature have created visions of an American past and future, establishing in the process a link between expansion and concerns for nature, and between nature and national identity.

CHAPTER ONE

THE WILD, WILD NORTH

Nature Writing, National Ecologies, and Alaska

[N]ature, like everything else we talk about, is first and foremost an artifact of language. As recent developments in literary criticism have made abundantly clear, language is not dependable. Trying to refer to things, language produces referents that tend to slip and disappear. As such, nature brought into political life, into language, can hardly remain "natural". . . . [A]ny attempt to invoke the name of nature—whether apologetic or confrontational in relation to authority—must now be either naïve or ironic. It can be anything but direct and literal.—William Chaloupka and R. McGreggor Cawley, "The Great Wild Hope"

Landscapes are culture before they are nature.
—Simon Schama, *Landscape and Memory*

Jon Krakauer's 1996 best-seller *Into the Wild* chronicles the experiences of Chris McCandless, the twenty-four-year-old nature enthusiast who left his suburban Washington, D.C., home in 1992 for a wilderness trek through Alaska's backcountry. Equipped with a ten-pound bag of rice, a small-caliber rifle, and not much else, McCandless fashioned himself into a modern-day American Adam determined to explore the nation's Last Frontier. The excursion came to an abrupt end four months later, however, when his emaciated corpse was discovered in an abandoned school bus not far from the boundaries of Denali National Park. Krakauer's account explains what drove the young man to embark on the adventure. By coming to Alaska, McCandless hoped to experience uncharted country, to locate an empty space on the map. Although it is doubtful whether blank spots existed anywhere in North America in 1992—or in 1492, for that matter—McCandless nevertheless devised a solution to his dilemma. Like a host of other American Adams before him, he managed to resituate Alaska as a terra incognita awaiting his arrival by simply getting "rid of the map."[1]

Between the time of its appearance in 1993 as an article in *Outside* magazine to its publication as a book in 1996, Krakauer's story elicited numerous responses, including many from Alaska residents who derided the author for glorifying what they saw as nothing more than a young man's folly. For these readers, McCandless represented just another ill-advised tenderfoot who ventured, unprepared, into dangerous country, hoping to discover answers to his life but finding instead "only mosquitoes and a lonely death." Over the years, scores of marginal characters have taken to Alaska's backcountry, many of them never to appear again, others living to tell their stories.[2] Accounts of adventurers surviving the wilds of Alaska have become such common fare, in fact, that Eric Heyne suggests we should consider Alaska literature as synonymous with tourist literature.[3] Shaped by the enthusiasm of the newcomer, this writing has done much to position the region as a unique terrain, the nation's last undeveloped wilderness area.[4]

Far from being the first American adventurer to seek untrammeled lands in the Far North, McCandless had important precursors in John Muir, naturalist and founding member of the Sierra Club, and Robert Marshall, mountaineer and founding member of the Wilderness Society. Both men wrote extensively about Alaska, participated in the mainstream environmental movement, and were largely responsible for popularizing the region as a limitless space of opportunity and adventure during the early twentieth century. As one of the most famous nature advocates to wander through the Far North, Muir helped establish standards of Alaska nature writing to the extent that nearly every writer following in his wake has had to come to terms with his literary constructions of the region. Over the years, Muir also has become a kind of folk hero in Alaska, his name still gracing several local sites on the state's map, including parks, trails, buildings, and glaciers. In a similar way, Robert Marshall has emerged as an important figure in the imaginative construction of Alaska. Based on his experiences in the Brooks Range during the 1930s, Marshall's books helped draw attention to this remote area of the nation's geography. By popularizing the region as a place offering an elusive sense of freedom not available elsewhere in the nation, the two men's writings have helped ensure that even today, Alaska is still understood through clichés that present it as a kind of blank slate or empty spot on the map. In many ways, these beliefs continue to shape how the dominant culture thinks and acts toward the region.

While Muir and Marshall provide early instances in which to examine popular understandings of Alaska, I am also interested in tracing their legacy in the writings of more recent figures. John McPhee and Barry Lopez produced some of the most widely read accounts of Alaska during and after the oil boom of the 1970s and 1980s, a period when a great deal of national attention became directed toward the region.[5] Both writers aim to recast depictions of Alaska as a Last Frontier, a place somehow set off and removed from the rest of the United States and the world. Restoring a critique of the political economy of nature writing to discussions of Alaska, McPhee and Lopez resist writing the region entirely within mainstream landscape conventions. Both of them rely on and extend conventional landscape rhetoric in their depictions of the region yet do so in order to challenge and critique the ways nature writers have historically imagined Alaska.

Drawing on the fields of cultural studies and ecocriticism, this chapter is largely interested in tracing the ways Alaska has become figured as a hypernatural landscape in the dominant American imagination and in examining the consequences involved in this kind of thinking. In order to argue these points, however, I must first pause to address the gap that has emerged between these two areas of study. While cultural studies and ecocriticism potentially have much to offer each other, their critical practices appear to be moving in divergent directions, with each field marked by a certain degree of suspicion toward the other. Although the area of cultural studies has shown an interest in developing environmental critiques—with Raymond Williams's *The Country and the City* (1973) helping to establish some of the groundwork for cultural studies as an intellectual practice while also setting the groundwork for a materialist analysis of landscape ideology—cultural studies of late seems largely disinterested in nature studies. Jhan Hochman traces the history of green criticism within cultural studies and argues that sustained forays into environmentalism have been "hampered by the discipline's guardedness against any mention of *nature*." As he points out, in cultural studies, "the terms *natural*, *naturalness*, and *naturalized* tend to be synonyms for *reified* and *essentialized*."[6]

If cultural studies has had a long and uneasy history with the concept of nature, then ecocriticism has likewise expressed wariness toward the central concerns of cultural studies. Interestingly, as ecocriticism in recent years has developed into a field of study in its own right, discussions of the political economy and the ideological unconscious of environmental writing have become increasingly marginalized. Much work in the field tends to see nature writing and environmental criticism primarily through an aesthetic or psychologizing framework, only to overlook the ideological and social aspects of narrative. Donald Pease warns about the recent "turn to nature" in cultural criticism, arguing that it has the potential to create a new academic tradition of "rugged individualism" that is as escapist and narrow as its cultural predecessor. For Pease, because studies of nature are never pure projects or ideologically innocent endeavors, they must always be "affiliated with other kinds of emancipatory literary practice" in order to have a progressive impact in the world.[7]

Ecocriticism often uncritically positions wilderness and nature writing as its primary objects of study and in the process celebrates nature as a restorative and regenerative force. In an extreme manifestation of this thinking, ecocriticism comes close to positing the idea that nature somehow remains beyond or outside ideology, entirely unmarked by the social. The more reductive version of this thinking abounds in caricatures of postmodern environmentalism and social constructionist theories of nature. The line usually goes something like, "You can't tell me that bear charging me in the woods is a *discursive construction*." Of course, the answer from cultural studies is that yes, even bears charging in the woods cannot escape being shaped by discourse, by the meaning systems different cultures map onto them. The degree to which the animal may be understood as, say, performing a maternal role in trying to safeguard her cubs, or as acting on predatory instincts at the expense of humans, indicates the ways even charging bears may be invested with meanings that shift and change according to various interpretive communities.

All of this points me to a larger tendency in ecocriticism with which I take issue here, namely, the desire to posit a transcendent understanding of nature, one that remains outside cultural demarcations. I would contend that in making arguments about "nature for nature's sake" or nature as a given, ecocritics miss the point about what is knowledge. Seeking to know nature from some objective point of view, a position unencumbered by the social, some ecocritics end up striving for an impossibility. Because of its interdisciplinary interests, ecocriticism often has borrowed from the natural sciences in addressing environmental literary concerns. In the wake of the Sokal/ *Social Text* debate, however, ecocriticism may have turned to science at an inopportune moment, at a time when poststructuralism and postmodern theories are being greeted with a certain degree of skepticism. Of course, there are many scientists themselves who reject the promise of objectivity, dismantling "the view from nowhere," or what Donna Haraway refers to as the "god trick."[8] Rather than holding out for a transcendent understanding of the natural, these scientists instead argue that knowledge always involves an intervention into the phenomenon, that the social cannot be so easily left behind.

Randall Roorda parts ways with many ecocritics by arguing that American nature writing has in fact operated as a profoundly social act. Noting that the process of retreat or what he calls "losing the humans" functions as a central dynamic in the genre, he recognizes that American nature writing is marked by certain contradictions. Roorda analyzes the persistence of nature writers who endlessly narrate the act of retreat to others upon their return. "What can it mean to turn away from other people, to evade all sign of them for purposes that exclude them by design, then turn back toward them in writing, reporting upon, accounting for, even recommending to them the condition of their absence?"[9] For Roorda, this evasion is a central aspect of nature writing; the need to "lose the humans" as well as the need to draw them back into the fold continually shapes our narratives of nature.[10] As his work implicitly indicates, the nature that these writers seek is also always a social entity, a nature imaginatively constituted for human purposes and pleasures.

While cultural studies may have set up the possibilities for studying nature in all its social manifestations, these potentials have not been fully realized. And while ecocriticism has an interest in interdisciplinary endeavors, the two fields have yet to converge in a satisfying way. In order for cultural studies to have a greater impact on environmental studies, scholars need to address their discomfort with nature and recognize that it is more than just a powerful element in processes of mystification. Likewise, ecocritics need to acknowledge that recognizing nature as a socially constituted entity is not an arrogant or egocentric concept, as many would like to believe. Instead, a realization of the ways language limits and constrains our understandings of the world may well help us avoid the assumption that science or even certain forms of literary criticism can have unmediated access to the real.[11]

The New World's "New World"

For the purposes of this study, I am interested in the ways Alaska has been imagined by nature writers, that group of individuals who traditionally offer their work as a step toward solving the environmental crisis. In the late nineteenth century, an American cultural discourse arose that foregrounded the vulnerability of the nation's natural landscapes. Faced with an ever-retreating wilderness and a seemingly closed frontier, nature writers responded by recovering, protecting, and exploring the last remaining wild places and blank spots on the map. John Muir was motivated by these desires when he traveled to the North in 1879. As one of the first cruise ship tourists to arrive in Southeast Alaska's Inside Passage, he wrote a series of travel accounts, one of which was used as a tourist brochure for the Northern Pacific Railroad and then revised for a larger audience. In this later account, Muir describes his experiences sailing through an area that was rapidly becoming a prime tourist location:

> To the lover of pure wilderness Alaska is one of the most wonderful countries in the world. No excursion that I know of may be made into any other American wilderness where so marvelous an abundance of noble, newborn scenery is so charmingly brought to view. . . . Never before this had I been embosomed in scenery so

hopelessly beyond description. To sketch picturesque bits, definitely bounded, is comparatively easy. . . . [B]ut in these coastal landscapes there is such indefinite, on-leading expansiveness, such a multitude of features without apparent redundance, their lines graduating delicately into one another in endless succession, while the whole is so fine, so tender, so ethereal, that all penwork seems hopelessly unavailing.[12]

In this travel account, Muir presents Alaska as an ineffable landscape, a terrain so overwhelming with possibilities that no attempt could be made to fully describe it. Using tropes of natural abundance, Muir constructs a vision of Alaska as a virgin land, an inexhaustible terrain, and an enticing entity. Unlike any region he has ever encountered, this landscape begs to be explored; it functions as an infinite space that actually takes an active role in "leading" visitors to its endless marvels. Muir's depiction situates Alaska as the New World's "new world," a Last Frontier that enables the United States to once again unmap and remap itself. According to the naturalist, Alaska constitutes a "whole country" in itself. It is an "expansive" and exceptional landscape that cannot be easily exhausted by the adventurer nor "bounded," he argues, by the pen.

Muir's descriptions offer a particularly interesting example of what Lawrence Buell calls the "not there" of nature writing, a principle of environmental representation that foregrounds the process of seeing rather than the object seen. Relying on cultural projection, this narrative strategy encourages the writer to become a "maker of views," fitting the landscape to his or her preconceived notions or aesthetics. For Buell, a central element in the "not there" of nature writing involves emptying and filling in the landscape, a process that encourages the belief that the nature writer is somehow discovering the region anew.[13] This rhetorical gesture functioned in an important capacity for Muir, who first traveled to the North only twelve years after the United States purchased Alaska from Russia. Popularly regarded at that time as a frozen wasteland, a region devoid of any social uses, Alaska became known alternatively as "Seward's Folly," "Walrussia," or "Iceburgia." Muir's descriptions, however, work to counter this view of the North, his celebratory language inventing Alaska as an important new scenic wonder, the nation's Last Frontier.[14]

Muir's attempts to situate Alaska as an exceptional landscape were aided by the particular geographical forces he encountered in the region. Throughout his excursions in Alaska, for instance, he expresses an interest in the area's frozen terrain, spending most of his time studying the "virgin mansions of the icy north" and the "shaping forces" of the glaciers. Fascinated by the power and action of the ice fields, he explains that "here . . . one easily learns that the world, though made, is yet being made."[15] Muir, especially attracted to this glacial region, proclaims that the area "was then unexplored and unknown except to Indians. Vancouver, who surveyed the coast nearly a hundred years ago, missed it altogether" (*Letters*, 47). Although accompanied and led by the Tlingit guides Toyette, Kadechan, Stickeen John, and Sitka Charlie, Muir presents this area of Alaska as an undiscovered terrain, emptying and filling in a previously explored land with his own words and meanings. Muir's traveling companion, the Reverend Samuel Hall Young, also engages in a similar project. As Young explains, "Almost the whole of the Alexandrian Archipelago, that great group of eleven hundred wooded islands that forms the southeastern cup-handle of Alaska, was at that time a *terra incognita*. The only seaman's chart of the region in existence was that made by the great English navigator, Vancouver."[16] Like Muir, Young also disregards Indian claims to and knowledge of the region; moving from Indian to British presence in the land, the missionary discredits both histories in order to position himself in the role of the region's first explorer.

This project of emptying and filling in the landscape surfaces again later when Muir and Young situate themselves as experts whose new knowledges enable them to improve upon earlier ones. During their visit to Southeast Alaska, for instance, the two men compare the sights they saw with what George Vancouver encountered during his visit to the region in the late eighteenth century. As Young explains,

> this region was changing more rapidly than, perhaps, any other part of the globe. Volcanic islands were being born out of the depths of the ocean; landslides were filling up the channels between the islands; tides and rivers were opening new passages and closing old ones; and, more than all, those mightiest tools of the great Engineer, the glaciers, were furrowing valleys . . . forming islands, promontories, and isthmuses, and by their recession letting the sea into deep and long fiords, forming great bays, inlets and passages, many of which did not exist in Vancouver's time. . . . Where Vancouver saw only a great crystal wall across the sea, we were to paddle for days up

a long and sinuous fiord; and where he saw one glacier, we were to find a dozen. (*Alaska Days*, 65–66)

According to Young, geological changes in the land enable the men to encounter more extensive terrain than the British explorer did, for the glacial transformations they confront open new passages that allow them to uncover additional landscapes. While elsewhere the possibilities for discovery appear to be at an end, the geological forces the two men encounter in Alaska grant them new adventures. For Muir and Young, the ice fields are not just exotic features of nature, but elements enabling the map to be endlessly reconstructed. With its glacial formations, this northern terrain constantly changes course; advancing and receding from one year to the next, the ice fields ensure that Alaska functions as an unmappable and thus an infinitely remappable terrain. Where other charts appear stable, foreclosing the possibility of discovery, the frozen and isolated region Muir and Young locate in Alaska remain open to new conquests. By presenting the area as a self-renewing terrain, the two men naturalized acts of conquest, depicting nature as a force offering itself up for inspection. In doing so, they helped relocate the American frontier to a northern landscape that promised to resist permanent acts of inscription. Unlike places that had to be imaginatively unmapped, this glacial world appeared to naturally unmap itself.

The shifting ice floes and the advancing or receding glaciers ultimately complicated the act of discovery, for areas previously charted by explorers often changed shape over time or even disappeared. Because earlier discoveries were frequently nullified by the geological forces of the ice fields, new charts of the region had to be constantly redrawn.[17] Attempting to capture the dynamic and shifting forces of nature, Muir portrays this glacial activity with excitement, believing it presents him with new opportunities for discovery. According to Young, Muir exclaims at one point: "We are going to write some history, my boy. . . . Think of the honor! We have been chosen to put some interesting people and some of Nature's grandest scenes on the page of human record and on the map. Hurry! We are daily losing the most important news of all the world" (*Alaska Days*, 67). Although the areas Muir "discovered" might not exist from one year to the next, he nevertheless presents himself as one of the fortunate few who were still able to uncover the mysteries of the world. Structuring his writings about the region as elaborate presentations of an expansionist drama, Muir himself explains that if staged properly, his accounts of Alaskan exploration might immortalize him and the region, placing both of them on the nation's maps.

Young illustrates a similar desire when he describes another trip he took with Muir in 1880. Sailing through the glacial fjords, Young explains the fascination he held toward the region: "Every passage between the islands was a corridor leading into a new and more enchanting room of Nature's great gallery" (*Alaska Days*, 71). The landscape Young encounters proves exciting because it appears to be unexplored and unknown; the terrain seems maze-like, its intricate pattern providing Young with the sense of adventure that seems unavailable elsewhere in the world. Contemplating the sights before him, Young also treats the landscape as if it were artwork on display in a museum, a valuable object in the natural world's "great gallery." Young's analogy functions as part of a larger preservationist rhetoric that presents natural landscapes as important national artifacts. Preservationists in fact struggled for legislation that would keep such places safe from development, safeguarding them for future generations. In 1906, Congress passed the Antiquities Act, giving the president power to name certain landscapes "national monuments," preserving them as "objects of historic or scientific interest."[18] The act of narration functioned in this capacity: Muir and Young's landscape descriptions helped transform the environment into an object of possession, allowing the two writers to invent both a specimen and trophy of their adventures. Like the exotic objects gathered from "primitive" places by the imperialist traveler, natural landscapes in faraway places such as Alaska also could be owned, preserved, and displayed in narrative form by the nature writer.[19] In their accounts of Alaska, the region thus emerges as an important new warehouse of raw material, with Muir and Young serving as collector and curator of the region's exotic specimen.[20]

During their travels, Young also employs an aesthetics of possession upon encountering an ice field not yet featured on their maps. He writes, "Without consulting me, Muir named this 'Young Glacier,' and right proud was I to see that name on the charts for the next ten years or more. . . . [B]ut later maps have a different name. Some ambitious young ensign on a surveying vessel, perhaps, stole my glacier, and later charts give it the name of Dawes. I have not found in the Alaskan statute books any penalty attached to the crime of stealing a glacier, but certainly it ought to be ranked as a felony of the first magnitude, the grandest of grand larcenies" (*Alaska Days*, 147). Young's playfulness here, his self-mocking outrage at being made a victim of theft, allows him to express his own complicated desires for ownership. Such desires betray his yearning for more stable projects of discovery, a sentiment that reverses his earlier sense of relief about the impermanence of Alaska's geographical boundaries. The problems of unmapping are complex, for the constantly changing terrain threatens to undermine Indian claims as well as Euro-American inscriptions in the North. Young thus realizes that his name, too, could be wiped off the face of the map by the shifting glaciers or by some later explorer.

During their Alaska travels, the two men continue to take imaginative possession of the terrain; while sailing in the region that would later become known as Glacier Bay, for instance, Muir turns to Young, who had earlier expressed a desire to visit Yosemite, and announces, "There is your Yosemite; only this one is on much the grander scale. . . . We'll call this Yosemite Bay—a bigger Yosemite, as Alaska is bigger than California" (*Alaska Days*, 156–57). According to John Jakle, place comparisons frequently serve to validate travelers' experiences by allowing them to link the sights they encounter with idealized landscapes that already have national reputations.[21] Muir's comments function in a similar way: his comparisons highlight the intertextual function of U.S. landscapes by foregrounding Yosemite as a reference point in generating cultural meanings for other regions. Muir himself was a central figure in the struggle to protect Yosemite, with the battle between him and the chief of the Forest Service, Gifford Pinchot, over the damming of Hetch Hetchy in Yosemite widely regarded as a turning point in environmental history. The conflict split the conservation movement, with preservationists such as Muir fighting to protect the region from development, and conservationists such as Pinchot supporting the construction of a reservoir and dam to provide San Francisco with a new source of water and power.[22]

Although Muir eventually lost the battle when the bill authorizing the Hetch Hetchy dam was passed in 1913, his struggle to protect Yosemite has led many historians to consider him a radical alternative to mainstream figures such as Pinchot, with some environmentalists arguing that while other nature advocates in the period emphasized the efficient and rational uses of natural resources in the service of industry, Muir fought this "materialistic resource management."[23] By positioning him outside the history of development, however, these scholars fail to recognize the ways he also campaigned for the careful management of nature through a preservationist ethics that involved managing the "scenic resources" of a region, a project he extended to Alaska. During his excursions among the ice fields, for instance, Muir positions himself as the quintessential resource manager by "offering" Alaska's "Yosemite Bay" as a gift to his companion.

While Yosemite initially functions for Muir as his most beloved landscape, he later endows Alaska's scenic resources with an even greater value than the California landmark. After sailing through the fjords of Glacier Bay, for instance, he insists that Alaska surpasses California, a place once celebrated as a site of promise and possibility. For Muir, the northern frontier replaced earlier frontiers because it seemed to provide more distinguished possibilities for discovery. In 1888, after returning to his home in Martinez, California, Muir lamented his present situation, telling Young, "I, who have breathed the mountain air—who have really lived a life of freedom—condemned to penal servitude. . . . I want to see what is going on. . . . So many great events are happening, and I'm not there to see them" (*Alaska Days*, 206–7). For Muir, California had become a mundane and exhausted land, a place associated with labor and toil.

In 1881, Muir returned to Alaska as part of the Corwin Expedition, whose members were charged with finding three whaling ships that had been trapped by the Arctic ice the same year. Muir invoked a language of display and ownership in describing his encounters with the land and the region's inhabitants. Walking a mile or so from the shore, the naturalist imagines how "fine it would be" to cut "a square of tundra sod of conventional picture size, frame it, and hang it among the paintings" on his study wall at home. Fascinated by the possibilities of this "Nature painting," Muir's thoughts are suddenly interrupted by the sounds of merry shouting. He looks up to find a "band of Eskimos—men, women, and children, loose and hairy like wild animals."[24] Commenting on what "a lively picture" they would make, Muir places the native inhabitants of the region within a preservationist rhetoric that situates "nature" as a crucial U.S. artifact and an important national possession.[25] His travel account thus engages in a process that Sylvia Rodriguez calls "transmutation," a form of mystification that involves imaginatively recasting a cultural space and its people into a nonthreatening version of what is actually encountered.[26] By describing the Eskimos as potentially valuable art pieces, Muir transforms them from their status as rightful claimants to and possible competitors for the land into objects of nature ("hairy wild animals") and then into objects of possession ("lovely pictures").[27] In doing so, he helps draw Alaska into the national orbit, presenting the land and its native inhabitants as important new resources in the production of an ever-expanding national entity.

Muir's environmentalist vision of Alaska continued to serve larger nationalist functions. In 1899, during the same time that the United States was fighting to relocate its frontier overseas, he again visited Alaska, this time as a member of the Harriman Alaska Expedition. Originally organized as a hunting opportunity for the railroad magnate Edward H. Harriman, the voyage was eventually transformed into a surveying trip designed to assess the feasibility of building a transworld railroad linking North America with Europe. Although it produced numerous volumes of materials concerning the natural history of the region, the expedition was largely considered a reconnaissance mission, resulting in few actual discoveries. This "floating university," with its assemblage of scores of university professors, artists, and scientists, however, proved to be such a spectacle that one reporter referred to the expedition as "the invasion of Alaska."[28] The report was not entirely exaggerated, for, during their voyage to the region, members of the cruise stole totem poles and Indian artwork which they later donated to the Smithsonian and their home universities. When they arrived in Prince William Sound, the members also vied with one another to name the ice fields after their affiliated Ivy League schools. As a result, glaciers in Prince William Sound became graced with such names as Harvard, Yale, Bryn Mawr, Radcliffe, Smith, and Columbia.[29]

Muir's description of his travels parallels the ambivalent sentiments nineteenth-century Euro-American writers often expressed toward the land; even as they mourned the loss of natural spaces, they nevertheless continued to generate images that celebrated the nation's course of empire.[30] As a nature writer and a friend of the railroad magnate, Muir was also not exempt from this project. In his biography of Harriman, for instance, he elaborates on the benefits the developer brought to the region, employing a fascinating use of environmental language in the process. "Of all the great builders—the famous doers of things in this busy world—none that I know of more ably and manfully did his appointed work than my friend, Edward Henry Harriman," he wrote. Responsible for revitalizing some of the nation's most impressive railroad lines, Harriman "went about it naturally and unweariedly like glaciers making landscapes, cutting canyons through ridges, carrying off hills, laying rails and bridges over lakes and rivers, mountains and plains. . . . He seemed to regard the whole continent as his farm and all the people as partners, stirring millions of workers into useful action, plowing, sowing, irrigating, mining, building cities and factories, farms and homes."[31] Deploying a language of nature for a decidedly antienvironmental cause, Muir celebrates Harriman's industrial drive in Alaska, producing in the process a remarkable account of "corporate green" and its early history in the Far North.[32]

Muir's depictions of Alaska also functioned in this capacity, proving so compelling to readers that thousands of tourists followed his path in order to experience the sights he described so seductively in his nature essays. Muir, however, lamented the arrival of these travelers, complaining that they failed to adequately appreciate the natural sites before them. Expressing his disdain for tourists during his 1890 trip to Glacier Bay, he writes: "[The ship] arrived . . . with two hundred and thirty tourists. What a show they made with their ribbons and kodaks! All seemed happy and enthusiastic, though it was curious to see how promptly all of them ceased gazing when the dinner-bell rang, and how many turned from the great thundering world of ice to look curiously at the Indians that came alongside to sell trinkets" (*Travels*, 245). While he sought to recuperate the landscape from the tourist gaze, Muir's fear that the region might be overrun with visitors ultimately reenacts a central aspect of tourism ideology. According to James Buzard, overstatements about a "tourist mass" recall similar anxieties the nineteenth-century middle class held toward urban crowds, particularly their potential for inciting social unrest.[33] Muir's antitourist rhetoric thus takes on a common form as he continually tries to present himself as an individual set off from the masses. For Muir, the crowd threatens to lay bare the ways his own travels were shaped by modern ideologies and fears, and the ways his desires were not unique or individual but actually shared by others.[34]

Because tourism is often thought of as destroying the authenticity of natural landscapes, nearly every traveler at some time or another seeks to disassociate him or herself from the class of tourists. This disavowal is so predictable that John Frow suggests it operates as part of the ritual of travel itself, "built into almost every discourse and almost every practice of tourism."[35] Buzard likewise argues that the tourist might best be treated as a mythic figure, a rhetorical instrument that emerged in the modern era to help delineate class identities, creating in the process a false dichotomy between traveler and tourist.[36] As he explains, the word "tourist" made its first appearance in English in the late eighteenth century where it served as a synonym for

"traveler." The formation of modern travel, however, required a denigration of the tourist; and in the modern ideology of travel, the complaint centers around a notion that the tourist appears "unable or
unwilling to cast off the traces of a modernity" which threatens to
destroy the otherness of foreign places.[37] Muir's use of the tourist as a
rhetorical device allows him to express ambivalence about other peoples' failure to escape modernity; that he himself was shaped by the
same forces is something that goes unnoted in his writings. Muir's
own ideas about wilderness advocacy, however, were modern ones
fully shaped by a modern understanding of nature and culture as
oppositional forces. In addition, his ability to travel to these places
with relative ease was enabled by a modern economy of tourism, an
industry that began to flourish in the late nineteenth century with the
publication of Eliza Scidmore's 1885 tourbook, *Alaska: Its Southern
Coast and Sitkan Archipelago*, and later with the popularity of his own
writings about the region.

Like the class of tourists from whom he distanced himself, though,
Muir aimed to locate an unspoiled terrain, an authentic place of refuge, and a natural attraction that had not yet become exhausted by
other visitors. In doing so, he also sought to restage earlier acts of
exploration in order to situate himself as the actual discoverer of a
region. This project led him to erase Indian knowledge of the land,
leaving his Tlingit guides virtually uncredited in his accomplishments.
At the same time, Muir's ecological vision celebrated the virtues of
outdoor recreation only to equate productive work with nature's destruction. For Richard White, this distrust of a particular kind of labor
contributes to a tendency common among mainstream environmentalists to view human beings as outside of nature.[38] In this case, however, Muir's recreationalist view of Alaska, and in particular Glacier
Bay, functioned to situate himself *inside* and the Tlingits *outside* of
nature, thus extending the possessive gesture toward Alaska that is
expressed elsewhere in his writings.

Muir's accounts did much to draw environmental attention to
Alaska and helped direct other nature advocates to the Far North.
While his narratives are noteworthy for the ways they popularize a
certain kind of environmental thinking in the region, his ecological
vision also had unintended effects. Because mainstream environmentalists typically focus on his efforts to celebrate and protect Alaska's
wild spaces, they often fail to address his close ties to the forces of
Americanization, his links to the tourism industry, and his participa-

tion as a member of various state-sponsored expeditions. According to Marcy Darnovsky, this oversight is largely due to Muir's elevated status in histories of the Progressive-era environmental movement. In typical accounts of the movement, a strict division is erected between preservationists and conservationists, with preservationists seeking to protect valuable wilderness areas from development, and conservationists advocating the careful management of natural resources through regimented governmental planning.[39] The division generally favors preservationists for their foresight in protecting the nation's wilderness areas for future generations, while chastising conservationists for their close links to capitalism. In some ways, however, the dichotomy between the two movements needs to be reexamined. Although important differences exist between conservationists and preservationists, the two groups nevertheless held a common vision, as both constructed a commodified object in nature that could be marketed, packaged, and sold to a wide range of consumers.

While mainstream environmentalists criticize conservationism for its problematic links with capitalism, they sometimes fail to examine preservationism for its elitist ideologies involving land use, especially for its tendency to preserve wilderness only for the privileged. The devotion to wilderness regions often had disastrous consequences for American Indians who were frequently displaced from the areas chosen to be preserved. And if conservationists sought efficient ways of developing natural resources for capitalism, preservationists themselves constructed what Darnovsky calls an "asocial environmentalism" that foregrounded wilderness over other places as the primary landscape in need of protection and that defined wilderness as a space from which humans are absent by definition.[40]

In many ways, Muir perpetuated this kind of environmental thinking, insisting that his landscape aesthetic was located beyond such practices. Despite his protests, however, Muir's depictions of Alaska were ultimately not immune to the industrial forces that eventually developed these "natural landscapes" as tourist sites. Becoming almost required reading among travelers to Alaska, his writings about the North have played a central role in transforming the region into a valuable asset for the emerging tourism industry. Having grown up in Alaska and as a frequent visitor to the state today, I still find it a common place to see nature enthusiasts traveling through the Inside Passage on the state's ferry system with dog-eared copies of Muir's nature books. More important, however, the popularity of Muir's writings also set in motion events that would ultimately disrupt the Tlingit Indians' ties to the land. His celebratory narratives helped secure the creation of Glacier Bay National Monument in 1925, which forced the Tlingits in the nearby village of Hoonah into sacrificing their best hunting and fishing grounds to nature tourism.[41] Today Tlingits still battle the National Park Service for the right to engage in subsistence practices in Glacier Bay, the latest struggle involving indigenous efforts to gather gull eggs, which the Park Service currently prohibits.[42] In that sense, Muir's advocacy had, at best, mixed results, bringing greater environmental attention to the area but doing so with grave consequences for the region's native inhabitants.

Contrary to the wishes of Muir and other nature enthusiasts, wilderness advocacy has not always prevented the nation's wild areas from succumbing to economic development, and in fact, some of the attempts to draw support for such areas may have helped urge development by presenting these lands as objects worthy of aesthetic consumption.[43] While promoting an environmental discourse that celebrated the sanctuary of nature, many preservationists popularized a landscape aesthetic that unwittingly transformed wild spaces into sites that only the privileged could appreciate. Ironically, the area constituting Glacier Bay that Muir explored and made famous has since been transformed over the years into a highly regulated and restricted national monument. Today, the only nature enthusiasts who can travel in Glacier Bay are those who can afford fares on cruise ships or flights on chartered airplanes.

"A Fortunate Latecomer"

While Muir is known best in Alaska for his excursions through the ice fields of Southeast Alaska's Glacier Bay, Robert Marshall is widely remembered as a forester who battled endlessly to prevent the devastation of the nation's forests by the forces of industrialism. Marshall first visited Alaska in July 1929, spending three weeks exploring the Brooks Range, a region whose peaks had not yet been subjected to Euro-American naming. He went home to Baltimore just as the stock market crashed in 1929, but returned to Alaska a few months later. In all, he made four trips to the Far North, spending almost all the depression years in Alaska.[44] A devoted advocate of the region's recreational values, he was also a founding member of the Wilderness Society, an environmental organization designed to protect North American wilderness from destruction at all costs.[45]

Environmental historian Robert Gottlieb has described Marshall as a paradoxical figure in the mainstream environmental movement, a nature enthusiast who was as passionate about advocating for the poor and powerless as he was for wilderness regions.[46] While he feared the loss of the nation's wild, undeveloped lands and urged the formation of a new organization that could fight for the protection of wilderness areas, Marshall also believed that these areas should be available to all people, not just the elite few who typically benefited from preservationism.[47] According to Gottlieb, Marshall's environmental vision included providing transportation to national forests for low-income people, funding camps where underprivileged members of society could experience the outdoors, changing Forest Service policies that discriminated against minorities, and securing lands near urban centers for recreational purposes.[48] Marshall's ideas about social equity and environmental protection, however, were controversial within the mainstream environmental movement, which eventually distanced itself from Marshall, especially after he was the target of red-baiting in the late 1930s. When he died unexpectedly at the age of thirty-nine, Marshall's will divided his $1.5 million estate among several projects, including one project for trade unions and other social advocacy projects, one for civil liberties, and another for wilderness preservation. The leaders of the Wilderness Society, however, downplayed his social vision, focusing instead on his protectionist ideas. Gottlieb argues that over time, much of Marshall's original legacy was forgotten or subjected to reinterpretation, his ideas about social justice taking a back seat to his ideas about wilderness preservation.[49]

Given this history, it is interesting to trace how a progressive figure such as Marshall could nevertheless end up advocating what were ultimately antisocial ideas about nature and the wild. In making sense of this problem, we need to understand the ways his beliefs were part of a larger history and how his environmental vision was motivated in particular by larger ideologies of nature, especially by notions that wilderness functions as a retreat from the world of problems. Seeking sanctuary from the ravages of depression America, Marshall himself fled the continental United States, where, he argued, the "codes of society and the proprieties of civilization had killed the spontaneity of the frontier."[50] His entry into the primordial world of the Last Frontier enabled him to abandon "the routine fight for a living."[51] Marshall's tales of the Arctic wilderness paint a portrait of an idealized landscape, "where there was no unemployment, no starvation, no slums, no crowding, and no warfare," a place diametrically opposed to depression America, where "there was greater misery, worse unemployment, more starvation than ever before in the history of the country."[52] Depicting the charm of living out a Waldenesque existence, Marshall constructs a mythical utopian space that seemed far removed from the troubles of the rest of the world.

Marshall's interest in wilderness adventures began when, as a young boy, he spent time in the Adirondack Mountains trying to see how many peaks he could climb in one day. During one excursion, he hiked for nine hours, climbing fourteen peaks and ascending a total of 13,600 feet.[53] The adventures offered by mountaineering thus allowed Marshall to unmap and remap the world and provided him with fresh opportunities and new purposes for exploration.[54] After deciding that the mountain landscapes of the continental United States no longer provided adequate adventures, he turned his attention to the North. Explaining his attraction to Alaska, Marshall expressed a particular fondness for "blank spaces on the map," especially those "virgin places" he believed were still located in the North. Upon discovering, however, that "only two really large sections were left uncharted," Marshall quickly set off to explore these last untrammeled lands.[55] An anxiety of belatedness structures his book *Alaska Wilderness*; at one point, he confesses that "for a time I really felt that while life might still be pleasant, it could never be the great adventure it might have been if I had only been born in time to join the Lewis and Clark Expedition."[56] The exploration of Alaska's remote mountain ranges, however, promised to rectify Marshall's tardy birth, allowing him to enact the thrill of discovery that earlier U.S. adventurers once experienced.

Like Muir, Marshall also describes Alaska as a place rivaling the natural sites found in the continental United States. The "Arctic wilderness" he encounters surpasses even "the grandeur of Yosemite," a textual reference and aesthetic yardstick to which Alaska was frequently compared. According to Marshall, had Alaska been located in the continental United States, it would have likely experienced commercial exploitation, becoming a "celebrated scenic wonder in the world accessible to tourists."[57] This Last Frontier is particularly valuable to him because it is a "no man's land," a vast region he regards as virtually devoid of any cultural signs. As Marshall explains, even the Eskimos had only recently named some of the area's natural features.[58] Most important, this seemingly untrammeled land resists exhaustion; for Marshall, it appears to be "such an infinitude of barely scalable mountains, that a person could spend many summers tripping from this center and still have fresh territory to explore."[59]

Marshall's wilderness adventures in Alaska ultimately enabled him to adopt a rhetoric that alleviated the crisis of his belated birth. After journeying through the Brooks Range, for instance, he explains:

> Often, as when visiting Yosemite or Glacier Park or the Grand Canyon or Avalanche Lake or some other natural scenery of surpassing beauty, I had wished selfishly enough that I might have had the joy of being the first person to discover it. . . . I yearned for adventures comparable to those of Lewis and Clark. I realized that the field for geographical exploration was giving out, but kept hoping that one day I might have the opportunity for significant geographical discovery. And now I found myself here, at the very headwaters of one of the mightiest rivers of the north, with dozens of never-visited valleys and hundreds of unscaled summits still as virgin as during their Paleozoic creation.[60]

Marshall's excursions through the northern terrain allowed him to think of himself as a trail-blazer, a modern explorer of unknown regions in the great tradition of Lewis and Clark.[61] At the same time, however, he took care to explain that discovery needs to adopt new meanings in the modern era; the motivations for discovery had to be reinvented as modern exploration has become increasingly confined to remote areas of the world where "few would conceive of settling." Polar explorations and other modern expeditions proved disappointing because they do not make "the road of humanity as a whole the least bit happier." As he pointed out, "the net result of these activities is to make mankind a little poorer because when an exploration is made there is that much less possibility left in the world for others to experience."[62]

While traditional expeditions were justified because they advanced human knowledge about the world, twentieth-century discoveries no longer provide this function. According to Marshall, exploration thus needed to be reinvented as a personal feat, as an experience culminating in an individual's personal gain, "that most glorious of all pastimes, setting foot where no human being has ever trod before."[63] Although he presented his travels in Alaska as personally motivated journeys, his endeavors were not politically innocent acts but operated as events shaped by the modern ideology of adventure. According to Michael Nerlich, adventure has always been a key element in the history of Western economic development by advancing or preventing the rise of certain class interests.[64] The modern ideology of adventure, however, frequently denies the material base for adventure, inventing instead a belief in "the disinterested adventure for the sake of adventure."[65] Within this ideology, travel gains importance precisely by distancing itself from the political economy that enables it. In the modern era, travel is instead recognized and acknowledged as a personal experience rather than a larger class venture.[66]

Seeking to legitimate his wilderness experiences in Alaska, Marshall relied on scientific transformations in forestry to provide a rationale for his eco-adventures, viewing the field of study as a "disinterested" vocation motivated solely by the adventure of science. Throughout his career, Marshall held several positions with the Forest Service, an organization that seemed to present a more positive alternative to the development policies that the National Park Service advocated. During the early twentieth century, competition for land jurisdiction between the two national agencies prompted the Forest Service to promote outdoor recreation and wilderness preservation. While the Park Service advocated tourism, a regulated, regimented, and middle-class way of experiencing the outdoors, the Forest Service advocated a less expensive and working-class alternative in the form of wilderness adventures.[67] In its efforts to thwart the power of the Park Service, however, the Forest Service did not prevent the economic transformation of national lands. During this period, the Forest Service was beginning its policy of multiple use, a program based on an industrial model that maximized the profits from resource extraction.[68] While Marshall regarded the Forest Service as a welcome alternative to industrial development, it may not have been the benevolent organization he wished. Frieda Knobloch points out that from its very conception American forestry was supported by a growing federal interest in claiming forests for the state. These organizations operated as part of the territoriality of the state rather than as institutions designed to merely save us from the "spoilers" of the public lands. Efforts to manage the nation's resources, Knobloch argues, are not disinterested acts but instead highlight the state's desire to control every aspect of the forests that fall within its domain.[69]

Marshall, however, claimed that his affiliations with forestry could provide his wilderness excursions with a veneer of authority while furnishing him with a sense of adventure, generating possibilities for future unmappings by opening new frontiers of knowledge. As he suggested, foresters and other scientists were in a unique position as adventurers precisely because they possessed new means of burrowing into the secrets of the earth. Thus, to his great fortune, Marshall's belated arrival in the North could be overcome and alleviated by his status as a scientific expert, a forester whose training and knowledge enable him to delve more deeply into the mysteries of the land.

A. Starker Leopold's foreword to *Alaska Wilderness* explains both the importance of Marshall's experiences in the Far North and the significance of his tardy birth. Marshall's scientific abilities, we are told, enabled him to enact frontier adventures in the modern era. "The vanguard of American frontiersmen was composed largely of unlettered trappers, traders, and prospectors who pushed their way into the virgin continent and left a scant record, or none at all, of the wonders they saw," Leopold writes. "In interior Alaska . . . most of the territory was first explored by nameless sourdoughs whose thoughts and adventures died with them. Parts of this inhospitable land were not penetrated at all until long after the gold rush had subsided. Robert Marshall was one of the fortunate latecomers who found a great reach of Arctic wilderness to explore and whose literary talents leave us an exceptional chronicle of the event."[70] Readers learn that Marshall was a "fortunate latecomer" because his predecessors themselves left a scant record of their endeavors. Acts of discovery, according to this logic, cannot be regarded as noteworthy in and of themselves; instead, these stories must be properly recorded for the future. Marshall's "literary talents," his ability to translate adventures into texts that could be consumed by others, were important contributions to the world's historians. The forester's belatedness was therefore transformed into his and our good fortune, as the accomplishments of those nameless explorers whose adventures died with them were replaced with Marshall's later account of journeys through many of the same lands.

Marshall also countered his late arrival by contesting indigenous authority in the region, appropriating their signs at one point by renaming a river "The Kenunga," an Eskimo word for knife. As he sheepishly explained, this act put him "in the same shady class of nomenclators as the poet Charles Fenno Hoffman who, nearly a century before, had taken the Seneca word Tahawus and placed it on a mountain the Senecas had never seen."[71] While he may have felt uneasy, such acts enabled him to recuperate his authority as a modern explorer. Only by utilizing indigenous words and co-opting a native subject-position could he imagine himself as one of the region's first real explorers.[72]

At another point, Marshall devised an elaborate means of overcoming the crisis of exhaustion modern explorers faced. He explained, "I do not believe that one man can get any more pleasure looking over 10,000 square miles by airplane than he could by exploring 500 square miles on foot, yet in doing the former, he would be robbing nineteen people of the inestimable thrill of first exploration."[73] Using the language of conservationism, Marshall advocated the "efficient" use of uncharted regions; and like environmentalists who called for the wise management of the nation's natural resources, he proposed conserving "discovery" and "adventure" themselves so that future generations could have the opportunity to experience these activities. Accusing fellow explorers of robbing each other of such pleasures, Marshall sought to curb the wastefulness of greedy adventurers whose modern modes of travel placed future discoveries at risk. According to him, even in the Last Frontier, wilderness enthusiasts must protect discovery from overuse, lest they, too, squander the raw materials that enable these adventures to be played out indefinitely.

Like other wilderness advocates who set themselves off from tourists, however, Marshall never acknowledged the ways his experiences in Alaska contributed to the land's economic development. Such responses have long operated as an important trope among nature enthusiasts. As Raymond Williams explains, wilderness is often supposed to function as a site offering peace and refuge, a place that "contrasts with man, except presumably with the man looking at it."[74] Assuming this position of innocence, Marshall returned to Wiseman, Alaska, after the publication of *Arctic Village* (1933) only to discover that the town had become a popular destination for tourists introduced to the North through Marshall's own writings. Although interest in the region grew partly as a result of his Waldenesque account of life in Alaska, Marshall regarded these changes with dismay, noting that his beloved town was "no longer the isolated community, uniquely beyond the end of the world."[75]

"Floating to Heaven on a Sea of Oil"

If John Muir and Robert Marshall were instrumental in situating Alaska as a Last Frontier, a modern sacred site much like Henry David Thoreau's Walden, then John McPhee and Barry Lopez have struggled to rethink this legacy in the late twentieth century, restoring a discussion of political economy to the region by addressing Alaska's industrial development. John McPhee visited Alaska in the 1970s; his bestselling book *Coming into the Country* (1977) addresses the state's boom years when the discovery of oil in Prudhoe Bay profoundly altered Alaska's economy and land-use policies. Barry Lopez's *Arctic Dreams: Imagination and Desire in a Northern Landscape* (1986) also examines the region in terms of a rapidly shifting economy while taking care to place Alaska in a transnational context as one of many regions that constitute the Arctic. Like McPhee, Lopez dismantles notions of Alaska as a pastoral or wilderness retreat, a place somehow set off from the rest of the United States or the world. Instead, his writings insist that we come to terms with the ways economic decisions in other parts of the world have a profound impact on this seemingly isolated and removed area.

McPhee's account begins in the Brooks Range, the celebrated site of undegraded nature made popular by Marshall in the 1930s. If Alaska is widely understood as the nation's most pristine wilderness area, then the Brooks Range has become known as one of the most wild regions in the entire state. Throughout his book, McPhee takes care to distance himself from some of the more problematic aspects of the mythic Alaska that Muir and Marshall helped set in motion. Discovering early on that Alaska is not a timeless or unchanging pristine terrain, McPhee instead encounters a region in the throes of a recent economic boom, its entire population "floating to Heaven on a sea of oil."[76]

In the wake of oil strikes in Prudhoe Bay and the building of the TransAlaska Pipeline, the state's newfound wealth helped draw people to the state, laying to rest any ideas that the settlement and development of Alaska would somehow prove exceptional or different from the settlement and development of the rest of the United States. Mc-

Phee learns, for instance, that just outside Anchorage, the waters have long since been fished out and are now supplied with stock from a state hatchery. As one of his companions explains: "The myth is that in Alaska there's a fish on every cast, a moose behind every tree. But the fish and the moose aren't there. People go out with high expectations, and they're disappointed. To get to the headwaters of a river like this one takes a lot of money. The state needs to look to the budgets and desires of people who cannot afford to come to a place like this." Although most Alaskans are urbanites—and in fact nearly half the state resides in Anchorage—there still exists a sense among many residents that Alaska represents something truly different from the rest of the United States. Yet Alaska's most populous city (often referred to as "Los Anchorage" by other residents in the state, a telling name to be sure) experiences urban troubles that closely replicate life in the Lower 48. At one point, McPhee ironically muses that while Anchorage "might be a sorry town . . . it has the greatest out-of-town any town has ever had." The irony is that many residents never manage to escape the city; with only two roads leaving Anchorage, the highways are continually clogged with traffic as weekend recreationists make their way to the wilderness. Much of McPhee's book goes on to recount stories of newcomers who leave Alaska soon after they arrive, having failed to encounter the wilderness they dreamed of mostly because they lack the economic resources to travel to these places (9–10, 133, 134).

If urban life in Anchorage is rife with the many of the same problems facing other American cities, then the town of Fairbanks presents its own challenges. According to McPhee, "Fairbanks has more motor vehicles per capita than does Los Angeles, and as the cars toot and tap bumpers on the long crawls through the ice fog the warm gases of their exhaust fumes seem to stick in the air close around them—an especially pernicious, carcinogenic subarctic variety of smog" (106). In McPhee's account, the Last Frontier desperately needs an urban ecology. Thoreau's pastoral legacy, responsible for shaping so much of Euro-American self-understanding, has become an environmentally costly one in Alaska. Yet even as the state faces dire environmental problems, such concerns are not necessarily foremost in the minds of Alaska's residents. Instead, one of the biggest debates in the state at the time McPhee visited Alaska involved whether or not to move the capital city from Juneau to Willow, a debate in which economic concerns rather than environmental ones took center stage. As residents

The 800-mile-long TransAlaska Pipeline

of Juneau worry about their city becoming a ghost town should the state's capital move to Willow, McPhee notes that the boosters who bought up land near the proposed site await the arrival of the state government and the huge profits that are sure to follow.

While he spends much of his book dismantling myths of the Last Frontier, McPhee also recounts tales of individuals caught up in their own modern-day wilderness fantasies. At one point, he tells the story of Bob Waldrop, a sort of updated counterpart to Robert Marshall. Disgruntled with the government's recent attempts to map, grid, and organize the state's lands, Waldrop devises various schemes to ensure that Alaska stays wild: "Waldrop . . . does not want his Brooks Range any other way. He wants it imprecise. He wants to preserve its surprises. When he goes up nameless mountains and finds on their summits containers identifying someone or other as the first visiting conqueror, he puts the containers in his pack and hauls them out. If you say to him, 'You're altering history,' Waldrop says 'The people were altering history who put the register there' " (*Coming into the Country*, 30–31). Engaging in his own process of unmapping, Waldrop discovers an ingenious way of leaving the region unspoiled and unexplored.

Throughout his account, McPhee also demonstrates that Alaska is not the last refuge of independent frontier types, but in fact is populated with individuals who are quite clearly hooked to the global. The

Pipeline with sign stating, "PLEASE DON'T CLIMB ON THE PIPELINE"

last section of his book records the experiences of the state's new breed of pioneers who live in and around the town of Eagle. Although shaped by a code of rugged individualism, these folks are very much dependent on the outside world for their livelihood. While one homesteader orders everything in from Seattle, including hundred-pound sacks of corn, pinto beans, unground wheat, cases of vinegar, tomato paste, butter, and dried milk, others are connected to the rest of the world by satellite, eagerly awaiting their orders from East Coast companies such as L. L. Bean (188, 376, 380).

Although for the most part McPhee tries to distance himself from the stereotypical sentiments directed toward Alaska, at times he, too, is caught up in some of the exceptionalist rhetoric shaping the region. At one point, he contends that Alaska is "in no way an extension of what I've known before." For him, "Alaska runs off the edge of the imagination, with its tracklessness, its beyond-the-ridge-line surprises, its hundreds of millions of acres of wilderness—this so-called 'last frontier,' which is certainly all of that, yet for the most part is not a frontier at all but immemorial landscape in an all but unapproached state." At another point, McPhee suggests that Alaska is a "rugged, essentially uninvaded landscape covering tens of thousands of square miles—a place so vast and unpeopled that if anyone could figure out how to steal Italy, Alaska would be a place to hide it." Replicating the language

Pipeline with graffiti

Muir and Marshall employed in their nature writings, McPhee contributes to the notion that the region is somehow an anomolous space unlike any other American terrain, a place fully set off from any other U.S. region (57, 133, 271).

While McPhee tries to employ this rhetoric for environmental purposes, the notion of an exceptional Alaska has historically worked in rather complicated ways. In his study of public responses to the 1989 *Exxon Valdez* spill, for instance, Thomas Birkland addresses how the symbolic value of Alaska has long affected national sentiments toward the state. While the region has been considered a pristine, isolated place largely untouched by humans—a sentiment that has contributed to the growth of environmentalism in the state—Birkland argues that these images of Alaska have proven effective for other projects as well, particularly those associated with the oil industry. The notion of Alaska as a changeless wild region, for instance, was useful to oil companies who used the myth to promote the idea that oil exploration could be compatible with the preservation of the environment.[77] As Birkland explains, because national responses to disasters are typically linked to the sites where they take place, the emotional impact following the *Exxon Valdez* spill was greater than it would have been had it occurred in a less celebrated terrain.[78] Immediately after the spill, the media picked up on the myth of Alaska as an exceptional

Tourists at the Alyeska Pipeline Visitor Center outside Fairbanks, Alaska

geography, intensifying the emotional impact of the disaster and creating in the process a huge public relations nightmare for Exxon. The national response to the spill eventually created a series of events that helped secure oil legislation that had been under deadlock for fourteen years.[79] In particular, it led to the passage of the Oil Pollution Act of 1990, which provides for tougher penalties in the event of a spill and allocates more resources for dealing with cleanups.[80]

Although it helped unleash a national response in the wake of the *Exxon Valdez* disaster, which helped change oil-spill policies across the nation, the myth of Alaska in part also has attracted industry over the years. Along with the state's policies that govern Alaska's natural resources, the mystique of the Last Frontier helped make the region an especially inviting site for development projects of all kinds. Lopez has lived in Alaska and across the Arctic for nearly five years, and he contends that it is hard to travel in the North today and *not* encounter industrial development.[81] His book *Arctic Dreams* describes the effects of oil exploration in Prudhoe Bay, the subsequent building of the 800-mile TransAlaska pipeline, the emergence of base camps for oil exploration that have also sprung up in Canada's Melville Islands, the huge mining operations that have been developed for lead and zinc in northern Baffin Island, and the hundreds of miles of new roads built to service these industries (xxiv). His work is crucial to evaluate in a study of Alaska because it focuses not necessarily on the state itself but rather the Arctic, a transnational region of which Alaska is a part. Ultimately, Lopez's choice of geography serves to restore Alaska to a global ecology and history, further dismantling ideas of the state as a Last Frontier, a site removed or set off from the rest of the world. In the process, he effectively overturns the concept of wilderness as an environmental logic as such sites ultimately cannot exist as sovereign space in and of themselves.

Lopez points out that while industries operating in the Arctic reside in different countries, clearly their developments spill over national boundaries, affecting life outside their national jurisdictions. For him, the project of understanding regions as fixed or static entities is not useful in the Arctic. Lopez's work thus provides an interesting illustration of the problems facing environmentalism in the global era, a concern many critics have foregrounded. In his study of contemporary ecopolitics, for instance, Thom Kuehls examines the ways geopolitical spaces in the era of globalization are always "pathetically porous" and the ways political, economic, and environmental activity flow beyond and exceed traditional borders and demarcations.[82] Pointing to the case of Brazil, a nation that has come to symbolize the global environmental crisis, Kuehls argues that the importance of the country's vast rain forest extends far beyond the borders of the nation: the function of the Amazon forests, for instance, in producing oxygen, absorbing carbon, and regulating weather patterns around the globe, highlights precisely the ways in which the Amazon "does not lie (solely) within the territory of Brazil."[83] According to Kuehls, the work of ecopolitics demonstrates not just the effects of decisions in one sovereign territorial space on other sovereign territories; instead, ecopolitics explores how sovereign territorial spaces are constructed in the first place.[84] Kuehls's observations about the ways the global complicates our understandings of territories further problematizes traditional uses of wilderness ideology. Because wilderness discourses function to close off one region from another as if the areas could be treated as hermetically sealed spaces, wilderness conventions, according to critics like Kuehls, might actually stand in the way of global environmental politics.

Throughout his book, Lopez also remains interested in rethinking territorial constructions that privilege the nation. The book begins by placing the Arctic into a history of globalization, chronicling first the development of British and American whaling in the nineteenth cen-

tury and then moving on to other accounts of Euro-American exploration in the region. Because the search for the Northwest Passage and
attempts to locate the North Pole cut across national spaces, Lopez's
study focuses on restoring a discussion of transnational economies to
the natural history of the region. For Lopez, the history of environmentalism in the Arctic has typically involved a series of transnational
players. For instance, in the 1950s when the polar bear was threatened
with extinction, a multinational response ensued. Concerns for the
animal resulted in the creation of an international agreement for polar
bear management, which shared information and coordinated management programs for an animal that typically does not recognize
national boundaries, but instead "drifts between countries and oc-
cupies the high seas in its wanderings." Using various tracking de-
vices, scientists involved in this project discovered that some bears are
long-distance travelers. A polar bear tagged in Svalbard, for instance,
showed up a year later 2,000 miles to the southwest, while another
bear traveled a straight-line distance of 205 miles in two days. For
Lopez, it is no wonder that the Eskimos of northwest Greenland call
the polar bear "pisugtooq," meaning "the great wanderer" (*Arctic
Dreams*, 71, 82).

In this account, the Arctic cannot be understood as a final preserve
for those seeking sanctuary from the problems of the modern world;
instead, the Arctic may be understood as a perfect place to examine
the dilemmas facing contemporary life. As Lopez discovers during his
travels in the region, traditional nature aesthetics have not always
served useful ends in the North. By delineating certain areas as worthy
of our attention and protection, landscape conventions have often left
other regions like the Arctic vulnerable to development. According to
Lopez, "The prejudice we exercise against such landscapes, imagining
them to be primitive, stark, and pagan, bec[o]me sharply apparent. It
is in a place like this that we would unthinkingly store poisons or test
weapons, land like the deserts to which we once banished our heretics
and where we once loosed scapegoats with the burden of our transgressions" (228). Lopez's account, then, aims to move beyond the
confinements of wilderness ideology; by describing a space that truly
exceeds these cultural demarcations, he seeks to replace traditional
landscape conventions with ways of seeing that understand the per-
meable nature of all environments.

Yet the drawback in discussions of global environmentalism, as
critics have argued, is that this emerging consciousness may end up

serving as a tool for renewed colonialist projects, with more power-ful nations dictating ecologies to less powerful ones. Joni Seager, for instance, addresses what she calls "pious environmentalism," the in-creasingly strident proclamations from wealthy governments about the need for other nations to manage their resources for the "global good." For Seager, when "the governments of the rich world embrace the arguments of environmentalists, arguments that they have resisted at most every other turn, there is reason to be wary."[85] To some extent, environmental critics have demonstrated that such histories have al-ready been played out in projects such as the turn-of-the-century con-servation movement. They suggest that U.S. conservationism should be understood as part of a larger colonial environmental movement that often involved the management of other nations' natural re-sources.[86] Anna Lowenhaupt Tsing argues along these lines, explaining that the nineteenth century witnessed plenty of interest in conserva-tion, much of it involving lands "elsewhere" in the European colonial periphery. Drawing on the work of historian Richard Grove, Tsing explains:

> One suggestive fragment [of this history] involves John Muir's in-terest in . . . Alexander von Humboldt . . . whose writings on the severe environmental consequences of European deforestation in Latin America . . . did a great deal to spread ideas of conservation from one European periphery to others. Muir originally planned to follow Humboldt's route in Latin America, to see the grandeur and vulnerability of nature as Humboldt saw it. . . . but health and financial limitations shifted his itinerary instead to the closer-to-home California landscape. There Muir entered a part of the United States that had only recently been Mexican territory and that was still ridding itself of Native Americans to create safe and culturally cleansed territory for white settlement.[87]

While global environmentalism may be an important endeavor to consider for ecocritics, such concerns are not without problems. To-day, as Tsing notes, the new conservationist is understood as no longer "just a white American" but a "global citizen" for whom the entire planet is now considered home.[88] The problem in this new global vision is the problem of difference. Although contemporary nature writers like Lopez are moving beyond their environmental predeces-sors in rethinking artificial borders and boundaries in their narratives, we are still far from resolving real issues of power between and among populations and environments.

Furthermore, in the case of Alaska, the project of dismantling the myth of the Last Frontier needs to go hand in hand with a consideration of indigenous land rights. Native environmental sovereignty in Alaska aims to counter visions and uses of the land as a Last Frontier, a terrain offering untold adventures for Euro-Americans. Alaska Natives have produced their own counternarratives that foreground their responses to U.S. environmental projects and that address land-use policies from an indigenous perspective.[89] An acknowledgment of Alaska Native environmental concerns therefore requires that critics understand the ways in which the landscape conventions underpinning dominant understandings of Alaska cannot be separated from longstanding cultural myths, from ideas of territorial expansion, and from U.S. nation-building projects. Such a recognition might also help critics reconsider what constitutes nature writing in other cultural contexts.

One of the most important forums for indigenous debates about environmentalism in Alaska, for example, has not been the traditional nature essay per se, but a weekly news publication named *Tundra Times*, which developed as an organized means of fighting the Atomic Energy Commission's proposal to use Alaska as a testing ground for nuclear energy in the 1950s. Seeking to develop peacetime uses for atomic weapons, the AEC devised a program in 1957 aimed at creating a climate more favorable to the development and testing of atomic weapons. Because of the dangers of radiation, a sparsely populated remote area needed to be located; according to historian Dan O'Neill, atomic scientists considered either the moon or Alaska for its testing sites.[90] They eventually chose the latter, justifying their decision through claims that they could "save" Alaska from its stigma as the "refrigerator of the world" by making it into an important new test site.[91] In 1958, atomic scientists announced plans for "Project Chariot," which involved creating a deep-water harbor in northwest Alaska near Cape Thompson. The Eskimos living near the test site resisted such uses of the land, and in 1961 *Tundra Times* began publication, its premiere issue featuring articles on Project Chariot as well as other stories about native land rights.[92] During its first year, the newspaper helped generate resistance to the AEC's plans, and Project Chariot was eventually dismantled.

Alaska Native environmentalism has often included concepts and ideas foreign to dominant American nature writing; indigenous environmentalism in Alaska understands, for instance, that conquest and genocide are aspects of a postcontact ecosystem. Indigenous environmentalism also resists understanding the subject and agent of nature writing as a solitary individual in retreat and instead concerns itself with the collective community. A reenvisioned understanding of nature writing that incorporates an Alaska Native perspective must therefore come to terms with the way the lone or antisocial nature advocate is not a useful model for political action; an environmental politics that leaves behind the concepts of territorial encroachment and conquest is doomed to failure from an Alaska Native perspective.

Unmapping America

Precisely because Alaska is still regarded as a Last Frontier in the dominant American imagination, the region has not remained immune to the problems of development. Yet many wilderness enthusiasts continue to believe that Alaska can somehow remain separated from social and political events taking place in other parts of the world. In his introduction to a collection of Muir's nature essays, for instance, celebrated writer John Haines laments what he sees as Alaska's recent decline. "Whatever is most characteristic of contemporary urban America, whether it is fast foods, mini-malls, shoddy housing, or just plain noise and ugliness, you can find it here in Alaska, and all the more objectionable for the lack of any originality whatever, and for being imposed on a land that cries out for originality and distinction. As my friend, the poet Philip Levine, remarked on seeing Fairbanks last summer, 'This looks just like Fresno.' "[93] For Haines and other wilderness advocates, the worst crime committed against Alaska occurs when people behave toward the land the same way they behave toward other places in the Lower 48. To see Fairbanks as Fresno in this instance is to see an unplanned modern city with all the problems other urban centers face. The concerns are made more pressing because of the ways Fairbanks seems to betray the promise of its wilderness surroundings. As Haines contends, the issue for many nature advocates does not lie with the treatment other American spaces receive; instead, problems arise when wild regions such as Alaska become straddled with too many cities like Fairbanks and Anchorage. I would argue, however, that Alaska's status as a uniquely wild space, as a location that "cries out for originality and distinction" can only emerge with the belief that the nation's other landscapes are somehow irrevocably lost. In this sense, the region's role as the Last Frontier, as a radically other American terrain, serves primarily to overcome U.S. environmental anxieties by enabling the United States to once again unmap and remap itself.[94]

Laguna Pueblo writer Leslie Marmon Silko confronts the problem of landscape from a different perspective. "It is dangerous to designate some places sacred when all are sacred," she writes. "Such compromises imply that there is a hierarchy of value, with some places and some living beings not as important as others. No part of the earth is expendable; the earth is a whole that cannot be fragmented."[95] If we take seriously the dangers of assigning certain places special status over others, then the environmental challenge facing the nation in the new millenium might be to stop promoting Alaska as a sacred region. As long as we imagine we still have sanctuaries like Alaska left, we need not ever seriously worry about areas lacking what we consider to be aesthetic value or somehow appearing beyond redemption. If we begin to recognize that the ecological concerns of places like Fresno impact Alaska, becoming Alaska's problem in turn, then the project of advocating wilderness preserves or pastoral hideouts is ultimately a limited project. Once we come to realize the profound ways in which the global impinges on every square inch of the planet, then the model of Alaska as a wilderness remove, national rescue, or Last Frontier can no longer be promoted as an environmental logic.

CHAPTER TWO

BORDER FICTIONS

Frontier Adventure and the Literature
of U.S. Expansion in Canada

[A]dventure is the energizing myth of empire. . . . The
American adventure stories represented . . . the policies
and compromises, the punishments and rewards, and the
stresses and problems involved in advancing a frontier at the
expense of native populations and against natural obstacles.
To read the adventures was to prepare oneself to go west and
take part in the national work.
—Martin Green, *The Great American Adventure*

[T]he specifically American construction of the modern
wilderness idea . . . serves increasingly to justify what are
effectively imperialist interventions anywhere across the
globe.—Denis Cosgrove, "Habitable Earth: Wilderness,
Empire, and Race in America"

In 1902, the frontier chronicler Frank Norris wrote that having gone as far west as possible, "suddenly we have found that there is no longer any Frontier." According to him, U.S. nation builders "went at the wilderness as only the Anglo-Saxon can," until they found they had arrived at the shores of the Pacific. The frontier reopened, he argued, when a "gun was fired in the Bay of Manila, still further Westward."[1] Facing a national landscape that seemed devoid of opportunities for adventure, many U.S. writers followed this expansionist move and sought new terrain in which to play out their western dramas of expansion and conquest. While some turn-of-the-century writers reconstructed U.S. frontiers overseas, still others such as Jack London, Rex Beach, and James Oliver Curwood relocated their western settings in the Far North.[2] At a time when environmental awareness was gaining popularity in the United States, these writers often discovered it was not enough to merely restage their narratives in a new setting. Thus, many of them struggled to preserve frontier experiences by framing their expansionist adventures in a conservationist rhetoric. In these narratives, the desire for an untrammeled land— an Other to the settled spaces of the western United States—reconstructed heroic acts as the struggle to save the environment. As a result, in many of these texts wilderness advocacy emerged as a form of imperial adventure in its own right.[3]

In his oft-cited "frontier thesis," Frederick Jackson Turner argued that the western frontier figured centrally in the production of European American identity. Using a language of nature, he contended that the frontier experience was crucial to the United States; the "forest clearings" were the "seed plots" of a national self that provided an "expansive character to American life."[4] By the late nineteenth century, however, national advancement across the continental frontier no longer seemed an infinite possibility; the "forest clearings" once considered important for the production of national identity were now transformed into signs of its exhaustion. Turner thus expressed new concerns, explaining that the "national problem is no longer how to cut and burn away the vast screen of the dense and daunting forest; it is how to save and wisely use the remaining timber."[5] Facing a scarcity of U.S. resources, Turner found he could no longer discuss expansion without addressing environmental concerns. The decline of the western American landscape led him to advocate a conservationist ethic that complemented rather than replaced expansionism; by connecting the two projects, Turner presented both enterprises as central elements in the national mission.

As critics have pointed out, Turner's observations about the closing of the frontier should not be understood as a truth claim but as an elaborate narrative strategy that helped rationalize U.S. expansion. Because the frontier in American cultural discourse is never *not* receding but is thought to be continually threatened with extinction, the national narrative is shaped by a never-ending quest for new lands, the conditions enabling a renewal and extension of an American self across time and space.[6] Turner used this rhetoric to advocate U.S. expansion in Canada, a land seemingly empty, unclaimed, and full of promise. "If we turn to the Northern border," he wrote, "we see in progress, like a belated procession of our own history the spread of pioneers." The "American advance" is now "carried across the national border to the once lone plains" and the "desolate snows of the wild North Land."[7]

Frontier writers also adopted a similar rhetoric about lands located outside the nation's borders. In his numerous adventure stories about the Far North, for instance, Jack London describes the region as a vast "White Silence," one of the world's last untrammeled spaces. His writings typically portray the North as a shifting signifier that includes Alaska and the Canadian Yukon, places whose national identities are always fluid and contested. Depicting these areas as interchangeable terrain, London presents them as a single, unified territory whose undetermined borders and geopolitical location allow U.S. adventurers unrestricted expansion across "the top of the world." In doing so, he contributed to a growing American interest in the Far North. At the beginning of the twentieth century, U.S. desires for empire became increasingly directed toward the region; although still a vague geography for most Americans, the Far North in general and the Canadian Arctic in particular held great romantic appeal as some of the last "blank places" on the map.[8] Adventure narratives about U.S. polar expeditions as well as other excursions across the North functioned centrally in bringing this geography into national consciousness. London's stories of American heroes in the "vast, silent" region likewise advanced these larger sentiments, depicting the Far North as an important new terrain for U.S. frontier adventurers.

Often situated in and around Dawson, a place one historian has called an "American city on Canadian soil," London's narratives helped establish a subgenre of the Western that featured U.S. adventurers moving north to Alaska and Canada in search of wilderness experiences, the continuation of their national mission.[9] The transnational frontier London described thus served an important role in expansion, and operated as a crucial staging ground for territorial struggles between natives, Russians, Canadians, and (U.S.) Americans.[10] As a way of justifying U.S. encroachment in the region, London framed his frontier adventures in an environmental rhetoric, depicting his heroes as "ecological subjects" whose care for the northern environment supposedly sets them off from other adventurers.[11] By placing this movement across the border, London's adventure narratives helped situate Alaska and the Canadian Yukon as an important wilderness region in need of protection as well as a new domain for U.S. nation-building enterprises.

Although he was perhaps the most famous U.S. writer of the Far North, London was not alone in drawing the area into the national orbit. Other American authors including Rex Beach and James Oliver Curwood also helped situate the region as an extension of the nation, creating an imagined community for the United States in a new northern setting. This chapter examines frontier adventure narratives written by London, Beach, and Curwood, and traces how a concern for nature in their novels helped shape understandings of Alaska while also aiding U.S. nation-building projects in the Far North. Although investigations of U.S./Mexico borderlands are increasingly challenging the ways we conceptualize American literary studies, U.S./Canadian relations still remain largely undertheorized, at least among scholars south of the border. An examination of frontier novels, however, allows us to trace the ways Alaska—a seemingly remote space on the nation's map—functions to resolve larger concerns about region, race, and nature. As a gateway for adventurers seeking new thrills in new lands, the Last Frontier aided U.S. projects of territorial expansion across the North in general and the Canadian border in particular.

The Greening of American Expansion

Although liberal ecological rhetoric typically presents nature advocacy as an inherently benevolent project, it is important to acknowledge the ways environmental awareness has also served larger national functions. Perhaps nowhere is this national project more notable than in the struggle over wilderness areas. As Denis Cosgrove points out, the history in which wilderness advocacy emerged in the United States suggests a "closer link with imperialist, xenophobic, and racist features of American nationalism than many Americans would feel comfortable espousing today."[12] This context has much to do with the connections drawn between the demise of the frontier and the racial discourses that shaped U.S. responses to immigration in the early twentieth century. Historians have argued, for instance, that both the perceived closing of the continental frontier and the rise of new immigration were seen as contributing factors in the decline of Anglo Saxon hegemony. In an era when white Americans felt overwhelmed by the growing numbers of new immigrants from other parts of the world, a beleaguered masculinity sought to reestablish itself through wilderness experiences, the continuation of the United States' frontier saga into the twentieth century. Outdoors adventures emerged as one means of reinvigorating U.S. men by allowing them to test their strength and endurance against the challenges of the wilderness.[13] In this context, the Far North functioned as a site of white flight, a new frontier where Anglo Saxon males could reenact conquest and reclaim their manliness.

A back-to-nature movement also arose during this period which advocated outdoors experiences as a crucial element of modern life. The movement had a wide urban constituency and included diverse interests such as hiking, gardening, scouting, nature study, mountaineering, and even voyages of discovery and Arctic exploration.[14] Although they lamented the unchecked development of the nation's natural resources, wilderness enthusiasts and back-to-nature advocates were typically not outside the very logic of development they criticized; by securing areas of the world as "wild," they themselves took part in transforming "natural" areas into social landscapes.[15] Moreover, in many instances, nature advocates also reenacted projects of conquest as the land they sought to preserve from development was often the home of native peoples.[16] Theodore Roosevelt as much as any other U.S. figure helps foreground the ties between nature advocacy and projects of expansion. As one of the nation's most famous wilderness enthusiasts and one of its most aggressive expansionists, Roosevelt argued that outdoors adventures created "masterful people" who contributed to the country's own virility. For him, the struggle in and for the environment operated as a new heroic adventure; the back-to-nature movement thus became both a means of promoting personal regeneration in the outdoors and a means of ensuring larger nation-building efforts.[17] By expanding into seemingly open spaces, U.S. adventurers could fulfill their nationalist destinies and help resolve the crisis of Anglo Saxonism. Rather than being opposed to acts of conquest, turn-of-the-century environmental discourses often functioned as the intellectual labor needed to ensure that expansionist projects remained an important element in national life.

In the late nineteenth century, western writers along with other adventure seekers traveled to the North—to Alaska and Canada—in search of wilderness opportunities that seemed largely curtailed in the continental United States. Jack London himself visited the Klondike in 1896 but returned to California in less than a year after contracting scurvy in the mining camps.[18] He stayed long enough in the region, however, to collect tales from miners about their gold rush experiences, information he later used as the basis for his short stories and novels. Known as the "Kipling of the Klondike," London did much to popularize the Far North as an important new frontier for U.S. readers, drawing this remote region into the nation's spatial imagination. What is interesting to notice about his tales of the Klondike, however, are the ways they typically erase Canada as a separate and distinct nation. In his discursive remappings of the North, London frequently depicts the Yukon and Alaska as one continuous territory, a primeval wilderness where adventuring U.S. miners struggle against the harsh conditions in order to prove their strength and survival as the fittest of all heroes.[19]

In "The God of His Fathers," for instance, London sets the story along the banks of the Yukon, a place where territorial conquest continues in all its "ancient brutality." British, Russian, and Indian presence in this multinational landscape eventually makes way for the arrival of the conquering Anglo Saxons from the South. The story opens with a description of this event:

> On every hand stretched the forest primeval. . . . Briton and Russian were still to overlap in the Land of the Rainbow's End. . . . The sparse aborigines still acknowledged the rule of their chiefs and medicine men, drove out bad spirits, burned their witches, fought their neighbors, and ate their enemies with a relish which spoke well of their bellies. . . . it was at the moment when the stone age was drawing to a close. Already, over unknown trails and chartless wildernesses, were the harbingers of the steel arriving—fair-faced, blue-eyed, indomitable men, incarnations of the unrest of their

race. . . . Like water seeping from some mighty reservoir, they trickled through the dark forests and mountain passes. . . . They came of a great breed. . . . So many an unsung wanderer fought his last and died under the cold fire of the aurora, as did his brothers in burning sand and reeking jungles, and as they shall continue to do till the fullness of time the destiny of their race be achieved.[20]

Through a strategy of unmapping, London presents the Far North as the destiny of Anglo Saxons from the United States whose white rule over the land is as assured as it is apparently elsewhere throughout the world. Portraying the struggle for conquest as a timeless U.S. activity, London uses organic tropes to describe American presence in the North, referring to the arrival of these men as a natural event, "like water" trickling through the wilderness. In this narrative, Euro-American expansion in the North emerges as a unique aspect of the national mission. The inevitable drive for U.S. rule throughout the continent brings new heights of civilization to this seemingly primitive land.

While western tales of this period typically featured the migration of an easterner whose own exhausted lands led them to move west, London's tales of expansion often chronicled the nation's expansionist drive to the North, and featured westerners who travel to Alaska and Canada in order to rejuvenate the United States' dying frontier ethic. Situating the North as an exotic new terrain in which to stage his local color literature, London once described the region as a "vast wilderness" encompassing "hundreds of thousands of square miles" that are "as dark and chartless as Darkest Africa."[21] Employing colonial motifs reminiscent of writers such as Rudyard Kipling and H. Rider Haggard, London conflated U.S. wilderness fantasies with British imperial desires in India and Africa, using a similar setting and plot in a manner that did not go unnoticed by audiences.[22] In a review of a collection of London's Northland tales, for instance, one writer suggested that "what Kipling has done for India . . . Jack London has done for the Arctic."[23] Another reviewer went on to praise London for outdoing the famous British adventure writer, claiming that he "goes farther than Kipling could go. His life in the North is more primitive, more elemental, than Kipling's Indian jungle life could possibly be."[24] Other reviewers elaborated on the importance of London's new choice of setting, especially appreciating the exotic possibilities offered by the Northland stories. "Life within the Arctic circle is so far beyond the stretch of our imagination that we scarcely reckon it in as a part of our world," one reviewer explained.[25] Another reviewer commented with enthusiasm that the Far North of London's narratives "has succeeded the Far West as the haunt of the adventurer, and opened up a new field to the writer of short stories."[26]

Born and raised in Oakland, California, London once expressed nostalgia for the western landscapes of the nation's past, lamenting what he considered to be the decline of the "old frontier." He confessed, "I realize that much of California's romance is passing away, and I intend to see to it that I, at least, shall preserve as much of that romance as is possible for me."[27] As historian Kevin Starr points out, frontier narratives written before the 1848 gold rush often portray California as an idealized agrarian landscape, the "cutting edge of the American Dream" where frontier experiences continue to be forged and where pastoral opportunities associated with Jeffersonian agrarian democracy might still be enacted. By the time London was writing, however, the Golden State no longer seemed to offer such promise. As Starr argues, London's descriptions of the California landscape often foreground the region's demise through signs of its ecological decline; his stories frequently portray the state as an exhausted space littered with "broken fences, overgrown roads, untended grapevines, crumbling adobe barns, and deserted mine shafts."[28]

In his Northland adventure narratives, London contrasts the decline of California with the possibilities offered by the Far North, a theme that figures centrally in his novel *Smoke Bellew*. The text opens as the main character, Christopher Bellew, is commissioned by a California newspaper editor to write stories about San Francisco that contain "the real romance and glamour and color of the place."[29] The job turns out to be a thinly veiled challenge, for, as the editor confesses, the project of writing "real" adventure stories set in San Francisco has become an almost impossible task. While California "has always had a literature of her own," he explains, "she hasn't any now." The literary exhaustion facing California leads Bellew on a quest for narrative. After hearing news of the gold strike in the Klondike, he proclaims that "the days of '49 are over." Later, Bellew follows the scores of Euro-American adventurers on their way to the days of '98, thus reshaping the nation's east-west expansionist trajectory into a movement from south to north. As the character explains, the opportunities offered by the Yukon gold rush are too good to pass up; the stories that will soon emerge from the region, he argues, are sure "to be big" (1, 12–13).

London's most famous text, *The Call of the Wild*, also presents the Canadian North as an important wilderness area for U.S. adventurers. Buck, the canine hero of *The Call of the Wild*, travels between Alaska and the Yukon, enacting his own back-to-nature movement when he "goes native," leaving his human companion, John Thornton, for life with a pack of northern wolves. The text reverses the project of expansion; at one point, for instance, Buck is sold to two French Canadians, a "swarthy" figure named Perrault and a "black-faced giant" and "half-breed" named François, who enter the Southland and exploit U.S. resources in the form of animal labor.[30] As miners crisscross back and forth between the Yukon and Alaska in the story, London erases geographical distinctions, celebrating the primitive, uncultivated qualities of the Far North while imaginatively claiming Canadian terrain for the United States. The constantly shifting settings in the text have often confused U.S. critics who associate the narrative with Alaska; following in London's footsteps, they help contribute to a national habit of erasing distinctions between the United States and Canada.[31] Such instances of border confusion are more than mere geographical oversights on the part of U.S. critics and writers. Instead, they foreground the ways wilderness ideology itself operates in the service of domination: by presenting the Yukon as wild, uncultivated terrain, the discourse of wilderness in U.S. culture becomes an expansionist gesture that suspends territorial boundaries and national jurisdictions across a given space.

As the gold rush drives London's U.S. adventurers north to Canada, the characters also adopt a colonial attitude toward the region, moving through the land in hopes of amassing great wealth yet rarely staying in the region once they've made their fortunes. In his story "To the Man on Trail," for instance, London tells of U.S. miners situated north of the border who experience "vague yearnings for the sunnier pastures of the Southland, where life promised something more than a barren struggle with cold and death."[32] London's novel *Burning Daylight* also opens with a depiction of the Far North as a silent, dead world, sealed off from the rest of "civilization" until the main character Elam Harnish arrives and brings it to life.[33] At one point while staying in the Yukon, London's hero tells other pioneers that he won't leave the region for the "outside" until he gets his "pile." He later announces his plans to "farm gold," an act that positions him as the quintessential U.S. figure, the agrarian hero situated, this time, in the Canadian North (32, 100).

Harnish's experiences in the "wild" also provide him with an expansive American self; he is described, for instance, as "too much" a man, "magnificently strong," and "almost bursting with a splendid virility," a figure unable to stay confined within the boundaries of the self or, as it turns out, the boundaries of his nation.[34] Because of these adventures, few men know him by any other name than "Burning Daylight," a title given to him "because of his habit of routing his comrades out of their blankets with the complaint that daylight was burning." A trailblazer, nation builder, and U.S. prospector on Canadian soil, Harnish is presented as the first white man to cross the "bleak, uncharted vastness" of the Chilkoot Pass to the Klondike, and a pioneer who "had made history and geography." This "King of the Klondike" secures U.S. rule throughout the "vast and frozen" terrain, earning a reputation for saving civilization from peril in the North. "Passing along the streets of Dawson, all heads turned to follow him . . . scarcely taking their eyes from him as long as he remained in their range of vision. . . . He was the Burning Daylight of scores of wild adventures, the man who carried word to the ice-bound whaling fleet across the tundra wilderness to the Arctic Sea, who raced mail from Circle to Salt Water and back again in sixty days, who saved the whole Tanana tribe from perishing in the winter of '91." As a frontier hero, Harnish helps stave off the threats to U.S. nation building in the Far North. He aids commercial whaling activities, keeps communication facilities intact, and helps to rescue rather than destroy the native population. His enormous reputation penetrates the farthest reaches of the "white wilderness," where he becomes an archetypal masculine figure for other Euro-American adventurers to emulate (5, 30, 68, 108, 124, 125).

In order to secure authority in the region, Harnish erases Anglo-Canadian and First Nations claims to the land and, through racial cross dressing, literally takes the place of the land's original inhabitants. At one point, for instance, we are told that Harnish's appearance has an "Indian effect": his hair is straight and black, his face lean and slightly long, and his garb made of "soft-tanned moccasins of moosehide, beaded in Indian designs." Yet in spite of these "native" qualities, the character's inherent domination as an Anglo-Saxon adventurer soon emerges; endowed with a unique destiny to rule the continent,

Harnish and his conquering activities are accepted and even anticipated by the other characters in the novel. The Tagish guide Kama, for
instance, describes Harnish as a nation-building figure whose presence
overshadows all others in the region. Kama's "attitude toward Daylight
was worshipful. Stoical, taciturn, proud of his physical prowess, he
found all these qualities incarnated in his white companion. Here was
one that excelled in the things worth excelling in, a man-god . . . Kama
could not but worship. . . . No wonder the race of white men conquered, was his thought, when it bred men like this man. What chance
had the Indian against such a dogged, enduring breed?" In staging this
moment of racial envy, London establishes his own supremacy while
presenting the Indian as doomed to vanish from the land. Kama thus
greets the rise of U.S. rule in the Canadian North as an inevitable and
unregrettable event. A spectator who contemplates the superiority of
the Western "man-god," Kama watches as the Anglo-Saxon nation
builders from the south descend upon the region in a conquering
mode that necessarily contributes to his own demise (7–8, 44).

While London presents the downfall of indigenous rule as an unfortunate but inevitable result of U.S. encroachment in the region, he
responds differently to the demise of wilderness. As Harnish soon
learns, the frontier adventures that U.S. miners enact in the Canadian
Yukon threaten to destroy the landscape so central to his personal and
national rejuvenation. At one point in the novel, he gazes out over the
mining grounds of Eldorado Creek and Bonanza, and comes face to
face with the devastating ecological impact of the "stampeders":

> It was a scene of vast devastation. The hills, to their tops, had been
> shorn of trees, and their naked sides showed signs of goring and
> perforating that even the mantle of snow could not hide. . . . A
> blanket of smoke filled the valleys and turned the gray day to melan-
> choly twilight. Smoke arose from a thousand holes in the snow,
> where, deep down on bed-rock, in the frozen muck and gravel, men
> crept and scratched and dug, and even built more fires to break the
> grip of the frost. . . . The wreckage of the spring washing appeared
> everywhere—piles of sluice-boxes, sections of elevated flumes, huge
> water-wheels,—all of the debris of an army of gold-mad men. . . . He
> looked at the naked hills and realized the enormous wastage of
> wood that had taken place. (117)

Like an invading army, the "gold-mad men" plunder and pillage the
Yukon in search of their fortunes, leaving a ravaged landscape in their

wake. This sight shocks Harnish, who realizes that frontier adventures cannot be sustained in the region unless careful planning is implemented. Becoming perhaps one of the earliest characters in U.S. literary history to grapple with the possibilities of "corporate green," Harnish decides something must be done to curb this waste. According to him, the ecological devastation emerged from the inefficient mining practices of individuals who extract natural resources with little interest in the cumulative impact on the environment. As he explains, this method "was a gigantic inadequacy. Each worked for himself, and the result was chaos. . . . it cost one dollar to mine two dollars, and for every dollar taken out by their feverish, unthinking methods another dollar was left hopelessly in the earth" (118). Historians have recently documented similar scenes of environmental destruction in the region, pointing out that during their rush to the mine fields, careless stampeders set endless forest fires in the Yukon, depleted the region's wildlife, damaged nearby streams, and left huge tailings in their wake.[35]

In other writings, London also details the damage caused by miners who show little concern for treading lightly on the land. In his novel *Daughter of the Snows*, for instance, the main character Frona Welse feels "vaguely disturbed by the throbbing rush of gold-mad men" who cross Dyea on their way to the Canadian border.[36] On the grassy flat where she played as a child, "ten thousand men tramped ceaselessly up and down, grinding the tender herbage into the soil and mocking the stony silence. And just up the trail were ten thousand men who had passed by, and over the Chilcoot were ten thousand more." Frona Welse notes, too, that the miners cared little about their impact on the natural landscape: they "laughed at the old Dyea River and gored its banks deeper for the men who were to follow." On the other side of the pass in Canada, the men continued to trash the trails, leaving behind their "overthrown tents and caches." As they made their trek in an "endless string," the men formed a "black line across a dazzling stretch of ice," becoming fainter and smaller until they "squirmed and twisted like a column of ants" (17, 39).

In *Burning Daylight*, however, the main character tries to distinguish himself from other U.S. miners by halting the environmental damage. Using the discourses of efficiency and expertise so central to the Progressive-era conservation movement, Harnish advocates resource management as a means of solving the region's environmental problems.[37] This conservationist project authenticates his presence in the region, enabling him to become an exceptional figure in the Far North, no longer an enemy of nature but its advocate, someone who reconfigures the invasion of Canada as a green activity. By endowing his character with environmental expertise, London effectively re-shapes nation-building events. Rewriting a history of conquest, he transforms U.S. expansionist activities in Canada into an ecofriendly enterprise. In her recent study of environmental history and the West, Patricia Limerick addresses the problems that arose in the wake of western mining rushes in a very different manner. As she explains, mining rushes often had profoundly negative consequences for the people living in the region. In particular, these events

> created the maximum degree of friction with Indians and set in motion the process that would leave them displaced, removed, and relocated. It is hard to imagine a system that could create more in the way of troubles for Indians: the discovery of precious metals and the movement to exploit them . . . flung white Americans around the Western landscape, into Indian terrain, in a way that left few areas untouched, and also left few reasons in the minds of the prospectors and miners as to why they should restrain themselves and their ambitions until a better arrangement could be made with the na-tives. . . . White Americans in mining rushes were clearly, unmistak-ably newcomers. . . . The recentness of their own arrival[, however,] did not cause them a moment's hesitation when it came to claiming the status of the legitimate occupants, the people who had the right to claim and use the local resources and to exclude and brand as illegitimate and undeserving people of other nationalities.[38]

London's narrative largely sidesteps this problem of legitimacy. Com-bining environmental awareness with an appeal to the regulated capi-talist development of the region, the frontier hero in *Burning Daylight* gains a reputation by presenting U.S. rule in the Far North as an ecologically benevolent project. What sets this figure apart from other miners, London suggests, is his environmental vision, his foresight in noting that frontier adventures inevitably face exhaustion unless something is done to stop ecological devastation. According to this logic, the only way for the nation to curb economic catastrophe, the closing of the northern frontier, and the decline of Anglo-Saxon mas-culinity is to adopt a "green" lifestyle and a rhetoric of wilderness.

An American Leatherstocking in Canada

After the days of adventure begin to wane in the Far North, Harnish follows the lead of other U.S. miners who, "having made their strike . . . headed south for the States, taking a furlough from the grim Arctic battle." He leaves the frozen Northland and resettles in California to try his hand at a new "game," this time the world of high finance. As it turns out, Harnish's experiences in the Klondike enable him to "grubstake" his economic adventures in the Southland. Upon arriving in San Francisco, however, Harnish learns that his earlier fame had died out and that in "no blaze of glory" did he descend upon San Francisco. "Not only had he been forgotten, but the Klondike along with him. The world was interested in other things, and the Alaskan adventure, like the Spanish War, was an old story." The Klondike adventure—depicted here as an Alaskan event—appears in a league with other U.S. imperialist adventures; both the gold rush and the Spanish American War are forgotten as the nation moves on to a new chapter of its history. Rather than feeling dejected, Harnish becomes excited by his invisibility, for it indicates to him how much "bigger this new game" is, when a man such as himself with his fortune and history passes unnoticed in a crowd. Entering what he calls "another kind of wilderness," the character confronts the world of capitalist adventures where he undergoes various losses and gains but finally multiplies his wealth several times.[39] The urban wilderness adventures Harnish experiences in San Francisco, however, prove less satisfying than the endeavors he carried out in the Far North. After spending several years battling other capitalists, he begins to feel the toll the urban wilderness has taken on his health, his "city-rotted body" no longer serving as the glorious receptacle of Anglo-Saxon virility (111, 123, 188).

The back-to-nature theme with its dual rhetoric of expansionism and environmentalism helps reverse the character's decline, a resolution London introduces through the figure of Dede Mason, a stenographer in Harnish's office. A lover of the outdoors and avid reader of Kipling's poetry, Mason disapproves of Harnish's capitalist enterprises and soon educates him about the regenerative qualities offered by the Californian landscape. Leading him through the mountains and the valleys, she teaches him to enjoy the "virgin wild," a natural landscape where "[n]o axe had invaded, and the trees died only of old age and stress of winter storm" (187). The novel ends as the main character gives up his financial struggles for a more peaceful agrarian existence in Sonoma Valley. The conversion to the pastoral dream appears at the close of the novel, when Harnish accidentally discovers gold on his plot of land; quickly covering it up with dirt, the character renounces his former way of life one last time.

Elizabeth Cook-Lynn has criticized white writers for their obsession with telling tales of a fondly remembered colonial past. The conquest of Indian lands by European peoples, she observes, is still largely portrayed in U.S. history and literature "as a benign movement directed by God, a movement of moral courage and physical endurance, a victory for all humanity."[40] Cook-Lynn's comments about the framing of conquest are meaningful in discussions of London's Northland adventure narratives. The environmental rhetoric featured in his stories invariably situates U.S. expansion as a "kinder, gentler" activity. This language appears in his adventure narratives as the Anglo-Saxon hero emerges not as a destroyer of the natural terrain or invader of indigenous lands, but a friend of the environment, a helpful ecological adviser. Throughout his texts, a concern for nature becomes incorporated into a national sense of self, where it prolongs U.S. frontier activities into the twentieth century. London's environmental writings thus foreground the territorial uses of wilderness rhetoric as nature advocacy becomes an important weapon in sustaining U.S. nation-building projects throughout the North. Denis Cosgrove makes a similar argument about contemporary nature rhetoric, pointing to connections between today's wilderness advocacy and U.S. colonial interventions across the globe. As Cosgrove observes, the wilderness rhetoric expressed today, though seemingly benign, often uses "the same language of pioneering adventure" and is frequently located "in the same theaters of snow, desert, and jungle" as "their colonial forebears" in the nineteenth century.[41]

Appropriating the North

Throughout American literary history, London has been classified as a western writer, and his stories set in the wilds of Alaska and Canada have been largely understood as western narratives. This classification of London and his work as *western* appears mainly because the United States does not have a strong tradition of writing the North like Canada does. London's frontier narratives are thus interesting for the ways they borrow and further develop the idea of the North, an emerging icon of Canadian nationalism that appeared during the late nineteenth century as a means of establishing Canada's cultural differences from the United States.[42] While other non-Canadian writers also adopted the iconography of the North, London proved to be the most successful, his name so closely associated with the Klondike that the region is still commonly misrecognized as U.S. terrain.[43] His production of a "Northland Leatherstocking" points to the ways that frontier adventure involved northern as well as western lands.[44] Although U.S. expansion is still widely understood as a drive from east to west and north to south, London's stories remind us that the push also occurred in other geographical directions, toward the north through Alaska and from there east into Canada.

Throughout history, a long line of U.S. politicians have advanced a similar vision of expansion, endorsing an ideology of continentalism, the drive for U.S. rule from the "Arctic to the Tropics." Continentalism largely influenced the foreign policy of William Henry Seward, the secretary of state who negotiated the 1867 purchase of Alaska from Russia. In a speech he delivered in 1846, for instance, Seward contended that the United States was "destined to roll its resistless waves to the icy barrier of the North,"[45] and claimed that the "enterprising and ambitious people" in Canada were building "excellent states to be admitted into the American union."[46] As secretary of state, Seward later secured the purchase of Alaska in the hope that the rest of Canada would follow suit and join the United States. An emerging Canadian nationalism, however, ended the northward march of the United States. During the same year that the United States purchased Alaska, Canadian leaders bought out the Hudson's Bay Company and annexed British Columbia in order to stop U.S. encroachment in the region. Although they faced new obstacles, many U.S. leaders continued to express desires for continental expansion, and even after the 1860s fears emerged north of the border that the United States might claim areas in western and northern Canada.[47] These anxieties have not abated in the present era, as the United States continues to extend itself onto Canadian soil in new ways; a postwar transnational highway, environmental pollution, polar expeditions, military installations, free trade agreements, and various other forms of cultural imperialism represent only a few instances of more recent U.S. border violations in Canada.[48]

Jack London was not the only U.S. writer to popularize the Far North as a site for frontier adventures or to depict expansion into the region as a rightful and benevolent activity. Rex Beach has also been identified as a northern writer; his novel *The Spoilers* (1905), filmed five times beginning in the silent era, was loosely based on his experiences in the northern gold rushes at the turn of the last century.[49] The environmental narrative that London used to frame his adventure narratives also appears in Beach's novel *The World in His Arms*. Set in the 1860s in San Francisco and Sitka, the capital city of Russian America, the novel reminds readers that the region comprising present-day Alaska functioned at one time as a multinational frontier with Russian, Spanish, British, and U.S. commercial interests competing in the area. Even as the novel makes use of this history, however, Beach portrays the space as rightfully American. His protagonist, Jonathan Clark, is a Boston fur hunter who raids Russian waters in search of endangered sea otters. The character falls in love with Marina Sclanova, the niece of the Russian governor, and then spends the rest of the novel convincing her to marry him. Clark also devises a clever scheme to enable U.S. expansion in the North, employing conservationist arguments that eventually convince the Russian government to turn the land over to the Americans.

Beach sets up this turn of events by describing the slaughter of seals in the U.S. fur industry. "It was bloody, heartless work which none of the white men enjoyed in the least. As a matter of fact, they hated it," he writes. "Jonathan Clark, for one, considered this wholesale destruction of harmless and bewildered creatures a thoroughly dirty and degrading business. He was ready to wash his hands of it in more ways than one."[50] As a poacher on foreign soil, Clark is an unlikely figure to espouse ideas of ecology in this context, yet he somehow makes a convincing argument to the governor after he is caught raiding Russian waters. "Your country took Alaska for the sea otters. They're gone, and the seals, which constitute the principal remaining source of revenue, are going the same way," Clark tells him. "You probably won't believe that a man of my sort can have a respect—a reverence, I may say—for the wonders of nature. But a rogue can revere beauty or grandeur and resent their destruction. Those fur seals are miraculous; it's a sacrilege to destroy them" (121). Persuading the governor to rethink Russia's territoriality, Clark sets in motion a series of events that enables the United States to purchase Alaska, a move imagined in the story as motivated as much by ecological concerns as by economic ones.

Throughout his career, Beach was a prolific writer of Westerns, with many of his stories published in the mass-circulation magazines that were a common venue for the genre in the early twentieth century. Like other short-story Westerns of the period, his narratives typically feature a male society in which women occupy minor roles. Set in a picturesque landscape always distinct from the urbanized East, the stories tended toward melodrama, employing over-the-top humor and emphasizing explosive physical action between the male characters who are presented as ideal frontier types, eager for adventure and intrigue.[51] After Glenister, the protaganist of *The Spoilers*, arrives in Nome, for instance, he announces, "This is my country. It's in my veins, this hunger for the North. I grow. I expand."[52] The description announces the arrival of a U.S. hero literally bursting with the urge for adventure. Like London's characters, Beach's expansionist heroes do not remain in Alaska; instead, their constant border crossings erase distinctions between Alaska and the Canadian Yukon as the characters struggle to take part in new outdoor adventures.

In a similar way, when Pierce Phillips, the protaganist in Beach's 1918 novel, *The Winds of Chance*, arrives at the international boundary on his way to the Klondike goldfields, he confronts the Canadian government's strict new mining regulations. "A ton of provisions and a thousand dollars!" the character exclaims. "Why that was absurd, out of all possible reason! It would bar the way to fully half this rushing army; it would turn men back at the very threshold of the golden North."[53] The "rushing army" or crowd of U.S. stampeders who follow in his wake share this frustration. Upon hearing the decree, they, too, voice their "indignation and bitter resentment." For Phillips, the Mounties' decision represents nothing less than the exercise of a "tyrannical power aimed at [his] ruin." Rather than becoming discouraged, however, he contends, "Most fellows would quit and go home but I shan't. I'm going to win out, somehow, for this is the real thing. This is Life, Adventure" (8–9).

The character's passionate insistence on his right to claim adventure in the Canadian Klondike captures the same sentiments London expressed in his narratives about the North. As Beach's hero explains, "Life and Adventure . . . that was what the gold-fields signified. . . .

He had set out to see them, to taste the flavor of the world, and there it lay—his world, at least—just out of reach. A fierce impatience, a hot resentment at that senseless restriction which chained him in his tracks, ran through the boy. What right had any one to stop him here at the very door, when just inside great things were happening? . . . a new land lay, a radiant land of promise, of mystery, and of fascination; Pierce vowed that he would not, could not, wait" (11–12). Staging its conflict as the struggle of an imperiled male body at risk—a U.S. subject denied its right to expand—the text illustrates the colonial attitude many U.S. writers adopted toward Canada. These territorial designs reemerged in the patriotic names U.S. miners later placed on the map. After the gold rush abated, for instance, many U.S. miners crossed back into Alaska, settling in places they named Eagle, Star City, Nation, American Creek, Washington Creek, and Fourth of July Creek.[54]

In *The Winds of Chance*, the U.S. adventure seeker is likewise presented as a forthright, patriotic character, an innocent figure wronged by a Canadian government intent on violating the pact of friendship and neighborliness supposedly negotiated by the two nations. Like London's writings, Beach's narrative foregrounds the emergence of a national subject who is blameless, virtuous, and upright, a figure who justifies the actions of U.S. adventurers on the other side of the border. National expansion seems inevitable precisely because of the persistence of this extraordinary frontier figure. Beach foregrounds this drama and adventure in a description of the rush to the Klondike.

> The . . . scenes of the great autumn stampede to Dawson were picturesque, for the rushing river was crowded with boats all racing with one another. . . . they went by ones and by twos, in groups and in flotillas, hourly the swifting current bore them along . . . Loud laughter, songs, yells of greeting and encouragement, ran back and forth; a triumphant joyfulness, a Jovian mirth, animated these men of brawn, for they had met the North and bested her. . . . they reveled in a new-found freedom. There was license in the air, for Adventure was afoot and the Unknown beckoned. (288)

While the international conflict of Beach's text highlights the struggle to assert a U.S. presence on Canadian soil, his novel also functions as a divided narrative. Like other Westerns that are shaped by what Forrest G. Robinson calls "bad faith"—the deliberate attempt to evade an unsatisfying evidence of truth—Beach's novel reveals certain elements of historical fact only to later counter that impulse with an act of concealment.[55] For instance, in foregrounding the "hateful redjacketed police" who prevent Pierce Phillips and other U.S. miners from claiming their rightful adventure in Canada, Beach's narrative reveals a larger pattern of denial surrounding their advancement across the border, and in doing so, enacts a double erasure of both Anglo-Canadian and First Nations presence in the region (41). Throughout the text, Anglo-Canadians emerge as the encroaching figures, while the Indian characters are reduced to their role as packers for the U.S. stampeders. Depicted as eager profit-seeking businessmen, the native packers are presented in the novel as unfairly reaping rewards from the suffering U.S. miners forced to secure a ton of provisions before they are allowed across the border into Canada.

The central plot of *The Winds of Chance* concerns the experiences of U.S. miners and their struggles with unscrupulous, greedy claim jumpers who cheat them out of their fortunes in the Canadian Yukon, a problem that becomes elevated to the level of national threat. At one point, for instance, Phillips faces a possible lynch mob after he is wrongly accused of stealing another miner's gold. He responds with amazement, explaining, "Something must be done. . . . It's tough to be—disgraced, to have a thing like this hanging over you. I wouldn't mind it half so much if I were up for murder or arson or any man's sized crime. Anything except *stealing!*" (403). Foregrounding the character's code of honor—his belief that theft is a worse crime than murder—Beach evades the central conflict of his novel, the problem that threatens to undermine the logic of his narrative. One could, in fact, locate the unspoken crime of Beach's text in the long history of U.S. territorial encroachment across the North, the uses of Alaska in American attempts to expand into Canada. Beach buries this nagging truth, however, and instead presents innocent, upstanding U.S. miners as the real victims of a theft.

The American resentment expressed toward Canadian officials during the gold rush that Beach features in his novel has been well documented by a variety of other writers in the period. In his 1897 guide to the Klondike gold fields, for instance, U.S. author Byron Andrews addressed similar border conflicts between American miners and Canadian officials, warning his readers that a "good deal of intemperate discussion" was beginning to emerge between the two parties. He claimed that "among other things a revolution has been proposed" with U.S. miners threatening to "proclaim themselves independent of the Canadian government; to erect a Territory of their own" to be called "Klondike," and eventually "raise the American flag" on Canadian soil.[56] Another U.S. writer, A. C. Harris, also described the nation's expansionist interests in the region, suggesting that during the gold rush, the Klondike had emerged as a "magic word that is thrilling the whole country. It stands for millions of gold and great fortunes for hundreds of miners, who have risen from poverty to affluence in the brief period of a few months." Captivated by the promise of an event so close to U.S. territory, he proclaimed that "[t]he old Spanish dreams of a wonderful realm somewhere in the Western Continent, made of gold and precious stones, seem almost on the point of being realized."[57] Describing the future of an American terrain in the Canadian North, Harris contended that an "interesting chapter" of history "is now in the making" and predicted that "in the near future the name of Lincoln will be given to a territory or state in the great northwest as that of Washington was some years ago" (277).

Such responses to the gold rush and desires toward territorial acquisition reveal how conceptions of U.S.-Canadian relations have been shaped in the American literary imagination and how the rhetoric of U.S. benevolence in Canada has produced its own fictions and fantasies. At the end of *The Winds of Chance*, Beach presents a fascinating resolution to the border conflict by staging Phillips's marriage to Josephine, the daughter of a Canadian mining official. The novel's conclusion functions as a thinly veiled wish fulfillment for a united relationship between Canada and the United States. The device allows the author to assert relations of power between the two nations as the feminized Canada becomes the subordinated party in the conventional tale of heterosexual union. First Nations claims, however, are erased in this marriage plot, which fails to script a role to third parties. In staging this political allegory, Beach's novel functions as part of an ongoing tradition aimed at absorbing Canadian claims of difference, independence, and sovereignty, and of overlooking an equally important indigenous history of displacement, dispossession, and erasure. In doing so, his narrative contributes to the fractious history that emerged as U.S. adventurers moved north into Canada in search of frontier experiences.

The arrival of American miners in the Klondike, in fact, had a profound impact on U.S.-Canadian relations and altered indigenous lifeways in the region in important ways. The boundary dispute that had been a source of international tension between the United States and Canada was eventually brought to a head as American opposition mounted against the 1825 treaty that had defined the boundaries of Russian and British possessions south of latitude 60. By 1898, however, both sides were increasingly interested in settling the controversy, especially as Canada's claims included Skagway and Dyea, towns that provided Canada with direct access to the sea without having to pass through U.S. territory. Recognizing they had much to lose, the Americans condemned Canada's claims, with Theodore Roosevelt refusing to allow the matter to rest. Fearing that the controversy would be settled to his nation's disadvantage, Roosevelt proposed that a tribunal of six members, three from each side, be constituted to assess the controversy. His U.S. delegates, not surprisingly, were less than neutral figures. In October 1903, Henry Cabot Lodge, Secretary of War Elihu Root, and Senator George Turner from Washington, a state that would be very much affected by the outcome of the decision about Alaska, helped sway the tribunal's vote in favor of the United States.[58]

Shortly after the decision was announced, a report in *The Dawson Daily News* indicated that, while Canadians opposed the outcome of the tribunal, they were not entirely surprised by the turn of events. The newspaper quoted one Canadian official who disapproved of the tribunal's decision to cede all territory to the states. "We wanted to live in peace and harmony with our neighbors, but it was time to call a halt. The states had established themselves to the west and the north. Was Canada to wait until she was entirely hemmed in by them?" Dismayed by this turn of events, he went on to predict that "[i]f the states discovered the north pole, then they would use it as a claim on Canadian territory."[59]

The U.S. and international response to the Canadian gold rush continues to have lasting effects in the history of the region, contributing to the imaginative production of the Yukon in ways that are often less than positive. The historian Ken Coates has pointed out, for instance, that in many ways the Yukon today is still unable to escape from what he calls "the limitations of an episodic history." The popularity of these adventure tales may be a determining factor for why the Klondike gold rush is still one of the few events in Canadian history that is widely known across the globe. Furthermore, while the U.S. onslaught in the Klondike displaced Canada's claims in the region, Canadian *Native* lifeways were also disrupted. Coates points out that the gold rush was almost overwhelmingly a nonnative event, as Indians in the region were continually placed at the margins of the action. Occasionally, Indians did participate directly in mining activity, staking claims on newly opened creeks but usually selling them at a huge profit to nonnative miners. Those Indians living near the vicinity of the mines were also involved in the rush but usually worked as packers, mine laborers, and traders, while Indians living farther away were involved primarily in trade with the whites.[60]

The marginalization of First Nations peoples is perhaps most telling in popular accounts of the Klondike's first gold strike. Conflicting national narratives continue to circulate across the continent as both countries argue whether the U.S. miner George Washington Carmack or the Canadian miner Robert Henderson made the first strike. The dominant version of this history credits the U.S. miner Carmack as the first miner to strike gold on Bonanza Creek, thus setting into motion the events that led to the Klondike stampede. It should come as no surprise to note that in London's novel, *Burning Daylight*, the author stages this national rivalry between Henderson and Carmack, with the American Carmack emerging as the Klondike's true hero.[61] This story is, of course, the U.S. version. In other accounts, the Canadian Robert Henderson is credited with making the discovery. In either case, these competing national narratives serve to secure both U.S. and Anglo-Canadian claims in the North while removing Indian presence from the region, for what is overlooked in these more popular accounts are the native actors.[62] While the U.S. version fits well with larger narrative conventions established by American adventure writers in the wake of the gold rush and the Anglo-Canadian version operates well as a national counternarrative, over the years yet another version of the story has emerged, this time from native oral traditions rather than from written ones. According to oral accounts, George Carmack's Indian companion, Skookum Jim, and Carmack's native wife, Shaaw Tláa (Kate Carmack), played a more central role in the discovery than George Carmack ever admitted. In the many versions he wrote about his experiences in the Yukon, however, Carmack always downplayed the role of his Indian companions, thus mirroring U.S. writers in the Far North who claimed the Klondike for their nation's history.

An American Promoter of the Far North

Like other frontier writers such as London and Beach, James Oliver Curwood also extended the United States' western experiences into the Far North by reimagining the Klondike gold rush as an American event, a project that likewise relied on an environmental rhetoric which allowed him to secure authority for his frontier heroes. Curwood was an active member of the U.S. conservation movement and eventually played an instrumental role in establishing legislation that preserved Superior National Forest in Michigan. Throughout his novels, he cultivated an awareness of nature, often expressing concerns that unchecked development would destroy the nation's "wild" landscapes. In his adventure narratives about the North, conservation operates not as a way of curtailing outdoor adventures but as a means of enabling and continuing the nation's frontier saga into the present era.

Curwood first gained fame for his stories set in the forests of the Midwest. His writings eventually caught the interests of the Canadian government, which offered him $1,800 a year plus expenses in 1901 to "explore the picturesque prairie provinces of the West." Curwood's task involved traveling throughout the Canadian prairies gathering materials for articles and stories that encouraged U.S. settlement in Canada, his career as an American promoter of the Canadian landscape resulting in over two dozen novels about an area he popularized as "God's Country." Curwood's adventure narratives were well read in the period; published in twelve languages, they sold more than four million hardcover copies in the United States alone. Over the years, Curwood came to regard his mission in Canada as an exciting experience, once confessing that "the opportunity to become a part of the great and glorious land whose far frontiers had been a part of my dreams for years thrilled me as no other event in my life."[63]

Curwood's interest in the Canadian landscape also led him to pursue Alaska as a setting for one of his frontier narratives. His 1923 novel, *The Alaskan: A Novel of the North*, for instance, is dedicated to "the strong-hearted men and women of Alaska, the new empire rising in the North." Like his other narratives, the novel functions as a thinly veiled promotional text advocating U.S. settlement in the Far North, an area of "immeasurable spaces into which civilization had not yet come with its clang and clamor."[64] As the author explains, while the North is a uniquely promising region, it is also not exempt from many of the problems other U.S. landscapes have faced. His novel thus begins with a depiction of the hero, aptly named Captain Rifle, who attempts to recapture the spirit of romance in an era when the boom of the Klondike had peaked and the mystique of the gold rush adventures in Skagway had died off. Set during a time when the region seemed threatened with the sense of exhaustion other U.S. regions previously faced, Curwood's text involves the struggle to preserve this endangered frontier.

The novel opens with a description of a die-hard U.S. adventurer whose name functions as a reminder of the nation's frontier past. "Captain Rifle . . . had not lost the spirit of his youth along with his years. Romance was not dead in him, and the fire which is built up of clean adventure and the association of strong men and a mighty country had not died out in his veins. He could still see the picturesque, feel the thrill of the unusual, and—at times—warm memories crowded upon him so closely that yesterday seemed today, and Alaska was young again, thrilling the world with her wild call." While the character associates the region with the promise and possibility of frontier adventure, he also notes a sense of sadness circulating among the residents who now seem nostalgic for the gold rush days. As he explains, "You can see it in their faces—always the memory of those days that are gone." Although the adventures that Alaska, and by extension the Canadian Klondike, once offered U.S. miners during the gold rush years appears a distant memory for the captain, he nevertheless tries to reclaim these glorious moments while seeking other means of rejuvenating the region (1, 6).

In order to promote Alaska as a site for present-day adventures, Curwood finds he has to counter dominant myths about the Far North. Although John Muir set in motion a different way of seeing the region that countered public perceptions of Alaska as a barren wasteland, a place devoid of any possible uses, this negative public sentiment was a force to be reckoned with even at the time Curwood was writing. In order to dismantle notions of Alaska as a god-forsaken terrain, he depicts his main character, Alan Holt, as an owner and manager of a reindeer herd, the northern equivalent of the American farmer. Curwood wasn't the first or the last U.S. figure to envision agrarian uses of Alaska. In an article published in the *Saturday Evening Post*, the chief of the Forest Service, Gifford Pinchot, also elaborated on Alaska's potential, telling of "the great stretches of agricultural land ready to produce in abundance the fruits, vegetables, and grains of Northern Europe," and the "cattle and the reindeer pastures, vast in extent," "the great resources of timber available for use in developing the mines," the general "fitness of this great land . . . to produce and support a population."[65] Such images promised to link Alaska to the rest of the United States, securing Euro-American dominion in the region. In his novel, Curwood helped stage these connections as well, in part by presenting Alaska Natives as employees on Holt's reindeer farm, an act that established them as part of an agrarian dream while dismantling their own claims to the land.

Addressing the region's status as a marginal U.S. territory in *The Alaskan*, Curwood highlighted the ways the land inflects larger themes about American national identity. His main character, Alan Holt, initially announces that his regional identity as an Alaskan holds more weight than his American identity. As an oldtimer in the region, Holt was "born in Alaska before Nome or Fairbanks or Dawson City were thought of." His love interest, who is aptly named Mary Standish, however, helps forge a more stable balance between Holt's regional and national identity. At one point after he foregrounds his regional interests over national ties, Standish interrupts him, crying out, "I am an American. I love America. I think I love it more than anything else in the world—more than my religion, even. . . . I love to think that I first came ashore in the *Mayflower*. That is why my name is Standish. And I just wanted to remind you that Alaska *is* America." Like his literary predecessor Owen Wister, who, in *The Virginian*, gave his New England heroine Molly Wood a revolutionary heritage that undergoes rebirth in the West, Curwood invokes a heroic American tradition in his heroine, Mary Standish, that is likewise rejuvenated by contact with the frontier.[66] Using images of the glorious founding of the nation to describe U.S. settlement in the North, he begins American history anew in Alaska by presenting the northern adventurers as pilgrims in a New World, an act that further establishes an American tradition in the North (5, 14).

Once Alaska becomes incorporated into the United States, however, the region risks the same threat of exhaustion that other frontiers already experienced. Holt realizes this problem and ponders the future of Alaska, wondering whether it, too, will follow the tradition of other U.S. landscapes.

> He looked out at the stars and smiled up at them, and his soul was filled with an unspoken thankfulness that he was not born too late. Another generation and there would be no last frontier. Twenty-five years more and the world would lie utterly in the shackles of science and invention and what the human race called progress. So God

had been good to him. He was helping to write the last page in that history . . . After him, there would be no frontiers. No more mysteries of unknown lands to solve. No more pioneering hazards to make. The earth would be tamed. (74)

While Holt feels fortunate to experience the last of the Last Frontier before it finally vanishes, he also realizes that the process of environmental decline is already unfolding before his eyes. The problem, as he explains, is that the once-promising gold fields have now become transformed into a popular tourist attraction. The steamships that earlier cruised through the Inside Passage to Skagway, taking U.S. miners to the start of their difficult trek to the Klondike, are now used as a "pleasure trip for flabby people" (16). According to Holt, this development signals an end to Alaska's status as a wild terrain. Tourism for him operates as a domesticated and feminized activity that transforms the region into an object of nostalgia rather than a manly terrain promising adventure and intrigue.

Holt's ambivalence toward the growing tourism industry stems from the ways it centers on older forms of adventure that the gold rush once offered. "Gold had its lure, its romance, its thrill. . . . It seemed to him the people he had met in the south had thought only of gold when they learned he was from Alaska. . . . It was gold that had been Alaska's doom. When people thought of it they visioned nothing beyond the old stampede days, the Chilkoot, White Horse, Dawson, and Circle City. Romance and glamor and the tragedies of dead men clung to their ribs" (42–43). According to him, the transformation of the gold-rush past into a modern tourist spectacle prevents Americans from realizing other opportunities offered by the region. His Alaska is thus endangered by a travel industry that has grown up unchecked in the short time since the end of the gold rush, by industrial development that remains unregulated, and by the ignorance with which most Americans still regard the region.

As a focal point for U.S. involvement in Alaska, however, the days of '98 also establish an important national tradition in the Far North. Thus, after criticizing the impact of tourism on the region, Holt himself becomes a travel guide for Standish, leading her around the town of Skagway and describing the extraordinary events that made the region famous. Pointing to the mountain range before them, he paints an imagined past for her, describing "the wind-racked cañon where Skagway grew from one tent to hundreds in a day, from hundreds to thousands in a week." Holt envisions "the old days of romance, adventure, and death" and describes for her the changes that were "creeping slowly over Alaska, the replacement of mountain trails by stage and automobile highways, the building of railroads, the growth of cities where tents had stood a few years before" (67–68). He guides his companion through the city, ensuring that she receives a proper account of Alaska's past, and in doing so, presents the region as a site once bursting with adventure, describing the passing of a great city that was now being replaced by the encroachment of "civilization."

The problem for Holt is that many Americans do not appreciate "the immeasurable space of the big country," and refuse to acknowledge the region's worth. At one point, he cries out, "What fool had given to it the name of *Barren Lands*? What idiots people were to lie about it that way on the maps!" (143). According to him, most Americans are still largely ignorant about the vast opportunities offered by the nation's Last Frontier.

> [P]eople don't know what they ought to know about Alaska. In school they teach us that it's an eternal icebox full of gold, and is head-quarters for Santa Claus. . . . Why . . . it's nine times as large as the state of Washington, twelve times as big as the state of New York, and we bought it from Russia for less than two cents an acre. If you put it down on the face of the United States, the city of Juneau would be in St. Augustine, Florida, and Unalaska would be in Los Angeles. That's how big it is, and the geographical center of our country isn't Omaha or Sioux City, but exactly San Francisco, California. (19)

The United States' acquisitionist history in Alaska becomes significant to Holt precisely because it enables the nation to expand its contours, allowing the country to once again redraw its borders. Presenting Alaska as an entity that is as important in the nation's development as the Louisiana Purchase was more than a hundred years before, he argues that the incorporation of Alaska reconstructs U.S. regional identities, making California part of the Midwest rather than the Far West. By situating California at the center of the country, the purchase of Alaska shifts the point of reference for the West, moving the frontier away from its present location. Alaska therefore helps the country enact a project of remapping by repositioning the geographical center of the nation from Omaha to San Francisco.

Yet if Alaska enables a national redefinition, its location also makes it a dangerous place for national security. At various points through-out the novel, for instance, the region emerges as a vulnerable frontier, an exposed land situated in close proximity to the Soviet Union. One character notes this danger, arguing that Alaska is "only thirty-seven miles from Bolshevik Siberia. . . . wireless messages are sent into Alaska by the Bolsheviks urging our people to rise against the Wash-ington government." Holt also expresses concern about the new crisis now facing Alaska, lamenting the danger of Bolshevism hanging over the region "like a smoldering cloud." As he explains, Bolshevism is "the menace of blackest Russia. A disease which, if it crosses the little neck of water and gets hold of Alaska, will shake the American conti-nent to bed-rock." The character thus makes conservation a key issue in the United States' national security, offering a solution that will allow Alaska to be developed carefully and profitably in order to guard against outside political influence. According to him, the real threat facing the region resides in the specter of its environmental and eco-nomical decline. Only through conservationism, the careful and effi-cient development of natural resources, can Alaskans be saved from the Bolshevik danger (19, 128).

Frontier adventures are therefore recast in the novel; rather than indiscriminately advancing northward, his heroes work instead to protect the land from selfish individuals whose careless acts threaten to destroy future adventures. Yet the frontier endeavors these charac-ters advocate also require a certain kind of environmental vision. As Curwood's hero explains,

> We have ten times the wealth of California. We can care for a mil-lion people easily. But bad politics and bad judgment both here in Alaska and at Washington won't let them come. With coal enough under our feet to last a thousand years, we are buying fuel from the States. We've got billions in copper and oil, but can't touch them. We should have some of the world's greatest manufacturing plants, but we can not, because everything up here is locked away from us. I repeat that isn't conservation. . . . And the salmon are going, like the buffalo of the plains. The destruction of the salmon shows what will happen to us if the bars are let down all at once to the financial banditti. (129)

Describing the collapse of the buffalo on the plains as a symbol of the closing of the continental frontier, Holt seeks to prevent a similar disaster from occurring in Alaska, the Last Frontier. The region is best served, he argues, by conservationist management, the careful development of natural resources rather than preservationism, an activity that "locks up" the land.[67]

Ecological management of this sort promises to secure Alaska from both Bolshevik encroachment and environmental doom. Expressing trust in Roosevelt's conservationist agenda, Holt claims that while some citizens criticize the president for "putting what they called the 'conservation shackles' on their country . . . he, for one, did not." (43). As he explains, "Roosevelt's far-sightedness had kept the body-snatchers at bay, and because he had foreseen what money-power and greed would do, Alaska was not entirely stripped today, but lay ready to serve with all her mighty resources the mother who had neglected her for a generation. But it was going to be a struggle, this opening up of a great land. It must be done resourcefully and with intelligence" (43–44). During the Roosevelt administration, the monopoly of land was a constant theme in national life, as large corporations threatened to exhaust the nation's natural resources. In an effort to prevent corporate control, conservation policy placed restrictions on private ownership, an act some westerners regarded as a "lock up" of the land.[68] In Curwood's novel, however, the main character suggests that Roosevelt's environmental policy is needed precisely in order to extend frontier adventures and ensure the future of Alaska as a viable American terrain. This vision is defined against preservationism, which, as Holt believes, closes off the land, in favor of conservation, which provides an "efficient and wise" use of the region's natural resources.[69]

Connecting ideologies of conservationism with expansionism, national security, and U.S. identity, Curwood's frontier hero argues that Alaska needs to be shaped by a foreign policy that situates nature as a crucial element in national security. Because uncontrolled industrial development threatens to turn U.S. citizens against the American way of life, he presents environmental politics as a constituting element of American identity and national security. Alaska's role in national security is also important for other reasons. According to the main character, the region is a crucial new territory, offering a jumping-off point for future expansion—not in the West or the North, but this time in the East. Holt elaborates on this idea, explaining that just across the Bering Sea lies Siberia, "the last and the greatest" frontier, where "not only men but nations would play their part in the breaking of it" (127). The novel's territorial vision thus links Curwood to London and Beach. Looking to the North for new frontiers, these three writers envision Alaska as a promising stepping stone that enables opportunities for U.S. expansion into other lands.

North to the West

In the stories and novels by early-twentieth-century frontier writers, the project of claiming new territory through Alaska helps recast expansion as an environmental act. The geographical oversights and the language of ecology that enable these authors to claim foreign land as U.S. terrain operate as part of a larger tendency among writers south of the border to blur, erase, or otherwise dismantle national boundaries. As James Doyle has argued, in the imaginative literature of the United States, Canada perennially figures "as a vague, peripheral, and ambiguous concept." Pointing to the U.S. habit of considering the northern country as merely an extension of itself, he suggests that Canada has always been an important element in developing an American national identity.[70] This use of an environmental rhetoric in nation-building projects becomes a means of greenwashing expansion, for in these narratives, U.S. encroachment in Canada is cast not for what it is, an act of territorial conquest, but as an ecofriendly gesture, a compassionate and charitable act. Discourses of nature thus emerge here much as they did in chapter one, as part of a larger expansionist gesture that operates centrally in the United States' national ecologies.

In an interesting turn of events in 1993 that might have unsettled popular frontier writers and their audiences, the right-wing Russian political leader Vladimir Zhirinovsky proposed that his country should likewise try to restore its previous "imperial greatness" by reclaiming Alaska, one of Russia's former American colonies. Noting the wealth of resources that Alaska has provided the United States since its purchase in 1867, Zhirinovsky argued that his nation should reextend its own frontier to include a significant portion of the Far North. Not surprisingly perhaps, the response among many Americans was disbelief that Zhirinovsky could suggest that Alaska still belonged to Russia, or that in some way ownership was contestable. In this instance, Americans were dismayed that another nation would attempt to make territorial designs on a region that was clearly a part of the United States.[71]

CHAPTER THREE

DOMESTIC ECOLOGIES AND THE MAKING OF WILDERNESS

White Women, Nature Writing, and Alaska

[H]umans in scientific cultures are placed in "nature" in gestures that absolve . . . unspoken transgressions, that relieve anxieties of separation and solitary isolation on a . . . planet . . . threatened by the consequences of its own history. . . . If history is what hurts, nature is what heals.
—Donna Haraway, *Primate Visions*

Solutions to environmental problems must be able to be imagined into the future rather than relegated to some idealized past.—Noël Sturgeon, *Ecofeminist Natures*

Often considered the nation's last great frontier, Alaska fascinates many Americans because of its status as a place set off from the settled spaces of the Lower 48. Alaska is largely considered to be wild terrain, the quintessential home of North American nature, a land that helps draw the United States' frontier past into the present era. Although it remains far from being undeveloped terrain, in the dominant geographical imagination of the United States, Alaska is nevertheless considered unsullied, unspoiled, and largely unmarked by culture. In this way, the region functions as "anachronistic space," Anne McClintock's term for geography figured as primitive or out of step with history, where time has somehow disappeared and where progress has long ago halted in its tracks. Anachronistic space for McClintock is land "perpetually out of time in modernity, marooned and historically abandoned."[1] Emerging as a colonial response that places cultural Others outside the Enlightenment time of the European self, anachronistic space figures geographical difference as a historical rupture. Alaska's location vis-à-vis the rest of the nation, its position as a northern terrain physically unlinked to the continental United States, allows it to exist in this "permanently anterior time."[2] The region's appeal lies in its position as antimodern space, its apparent ability to resist change and the ravages of history while remaining fully archaic.

As I've indicated in the previous chapters, Alaska's position in the U.S. spatial imagination also makes it an important locale for Euro-American men who seek the wild, that element of the national past which modernity promises to forever banish. Within this logic, Alaska is largely defined and understood as male space, a playground where white adventurers may flirt with the primitive while avoiding the cultural degeneracy they often project onto their geographical Others.[3] Although this study so far has focused on Euro-American male writers and their textual formulations of Alaska, I do not wish to imply that these figures are the only U.S. writers who have taken up Alaska as a point of interest for their narratives. In this chapter, I outline other voices that have both aligned themselves with and contested the project of imagining Alaska as a Last Frontier; I thus shift focus here in order to trace how white women writers also contributed to the myth of Alaska, even as they faced a space that was predominantly regarded as male.

By the late nineteenth century, white women were producing vast amounts of writing about the region, much of it appearing in books and popular press magazines in the form of travel accounts, settler narratives, and nature essays. With titles such as "I Was a Bride of the Arctic," "Honeymooning in Alaskan Wilds," "Woman in the Wilderness," and "I Teach French in Alaska," these writings recognized that Alaska was thought to be better suited to enterprising white men even as they also sought to carve out a space where white women could claim their own adventures.[4] These writings did so with varying degrees of success; in the late nineteenth century, for instance, the most popular and authoritative travel guide about the region was penned by Eliza Scidmore, whose descriptions of the Inside Passage became required reading for tourists cruising the waters of Southeast Alaska.[5] My point in foregrounding these accomplishments, however, is not merely to commend such endeavors. By resisting a celebratory impulse, I hope instead to connect white women's experiences within the literature of nature, travel, and adventure to the larger development of national ecologies in the Far North.

This chapter examines in particular the literary work of two white women nature writers in Alaska. In their accounts of the region, Lois Crisler and Margaret Murie described their attempts to forge a place for themselves in the wilds of the North, and in doing so, produced texts that played a crucial role in drawing Alaska into the nation's imagined community during the post–World War II period. Encountering the region during a time when it was undergoing enormous environmental changes, Crisler and Murie struggled to write a poetics of place in light of these transformations and sought to create new landscape conventions that would be appropriate to this rapidly changing terrain. I begin with a discussion of Lois Crisler, a former English instructor at the University of Washington whose autobiography, *Arctic Wild* (1956), was the first book published on the Brooks Range since Robert Marshall's accounts of the region appeared in the 1930s.[6] Crisler is widely recognized for her efforts to champion new wildlife management policies, her attempts to develop different understandings of predator species, and her struggles to end the eradication of wolf populations across North America. Crisler likewise contributed to the production of *White Wilderness* (1958), a popular Disney nature film that went on to win an Academy Award for best documentary in 1959.[7]

I then turn my attention to Margaret Murie, a figure who has been popularly regarded as the "mother" of the modern environmental movement.[8] Raised in Fairbanks during the early twentieth century,

Murie spent her childhood in Alaska, moved away to Oregon for two years while she studied at Reed College, and then came back to the North where she spent the early part of her marriage working in the field beside her husband, the preeminent wildlife biologist Olaus Murie. The Muries' six years of fieldwork eventually led to the creation of the Arctic National Wildlife Refuge.[9] The Muries later settled in Moose, Wyoming, but frequently returned to Alaska for various research trips and lobbying efforts over the years. After the death of her husband in 1963, Margaret Murie forged a career of her own as an environmental activist, traveling across the country on the lecture circuit promoting efforts to preserve the nation's last wild lands.

Although these two women writers were widely read by their peers, Crisler and Murie have been largely overlooked by contemporary scholars. This oversight, I argue, has much to do with the ways writing about Alaska is still associated with male figures such as John Muir or Jack London, and the ways nature and adventure writing in general are shaped by a master narrative involving an American Adam and his strategy of return to the wild.[10] In this chapter, I examine how Crisler and Murie managed to insert themselves into a space that typically excludes or marginalizes female experiences. In placing their work alongside that of their white male counterparts, however, I also challenge recent studies that adopt an overly celebratory response to white women's nature writings. Although I am interested in examining their struggles to narrate a different understanding of place and self, I intervene in larger discussions about gender, culture, and the environment by attending to the ways white women often managed to invent an authoritative self upon entering the wild and the ways they frequently used nature writing as a means of mediating cultural and geographical otherness.

Writing Women, Writing Nature

In April 1953 after being commissioned by Walt Disney to film wildlife for a nature documentary about the Far North, Lois Crisler set off with her husband, Herb, for an eighteen-month sojourn in the Alaskan Arctic. Although the Crislers had spent much of their married life making a home for themselves in the wild, Alaska represented a different challenge for the two of them. "People think of an expedition into remote wilderness as something like a big military affair planned long ahead by many helpers, or else as a safari with guides who know the country and the equipment it calls for," Crisler wrote in her account of the excursion.[11] As she soon discovered, however, their expedition to the North would be of a much different scale. Although they relied on supplies dropped off to them by plane, the two were otherwise very much on their own. Before boarding the small aircraft that was to transport them to the North, their bush pilot warned about the attractions and difficulties of the land they were to encounter. "[I]ts great freedom will haunt you with longing," he said. "You'll pray to get out and cry because you can't come back" (*Arctic Wild*, 1).

The Crislers also faced challenges of another sort. After several weeks passed and they were still having trouble obtaining the wildlife footage they needed for Disney, the two realized they had to rethink their approach to filmmaking. The wilderness world they thought was at their fingertips in Alaska proved less accessible than they had initially hoped, and it was all too easy to miss capturing that perfect shot. To rectify the situation, the two of them decided to take a different route by filming animals in captivity. In *Arctic Wild*, Lois Crisler relates her experiences living with domesticated wolves in Alaska. At a time when the practice of captivity was a subject of debate among American naturalists, Crisler's account betrays an uneasiness about her responsibilities toward the animals.[12] The text likewise expresses an ambivalence about boundary violations, about transgressing the borders that separate the wild from the civilized, the animal from the human, and the world out-of-doors from life domesticated.

In the post–World War II period, Disney nature films were an important venue for shaping ideas about the environment. Although the Disney company did not create the genre, its nature films departed from the Hollywood formula at the time by adopting a more subjective and sympathetic approach to nature. In doing so, the studio also helped develop a particular way of seeing the environment that still influences present-day nature films and television programs.[13] For many American audiences in the 1940s and 1950s, the natural world often provoked strong responses. Extensive environmental upheaval after the Second World War profoundly altered American communities as modernization reshaped the national landscape through the creation of new suburbs, freeway networks, and industrial agriculture. Responding to these social changes, many Americans made efforts to reconnect to the "natural" in more authentic and satisfying ways. The Disney films functioned in an important way to mediate this nostalgia for a bygone nature, helping Americans forge new relations with the environment by defining the proper relationship between human and nonhuman nature.[14] As Gregg Mitman argues, the Disney company heeded this desire by providing a genre of "sugar-coated" nature films that portrayed images of an innocent American past, a nature that had somehow not yet been compromised or significantly altered by the forces of civilization.[15]

Alaska and the Pacific Northwest eventually emerged as the settings most favored by Disney photographers seeking to capture pristine American landscapes. The company's first nature documentary was filmed in the Pribilof Islands of Alaska and featured the life cycle of seals in a film called *Seal Island*. Released in 1948, the movie inaugurated the studio's "True-Life Adventure" series and proved to be a tremendous success, winning the Academy Award for best two-reel feature documentary the following year. By the mid-fifties, the studio was releasing about one nature film a year, with subsequent movies such as *The Living Desert* (1953), *The Vanishing Prairie* (1954), and *White Wilderness* (1958) also garnering Academy Awards. Part of the success of these documentaries may be traced to the filmmakers' familiarity with and loyalty to Disney plot formulas. Because many of the early producers of the nature documentaries specialized in animation, where each frame of the film was subordinated to the storyline, Disney's success in the earlier animation films was easily translated into the nature documentaries. The filmmakers also used new techniques in slow-motion and time-lapse photography to create the effect they needed for good storytelling.[16] Although Disney emphasized the "true-life" quality of the films, the studio's nature documentaries actually provided instances of a highly staged and constructed nature. Walt Disney's own goals for the series involved capturing the entire natural history of an animal without including any sign of humans. Robert De Roos, a writer for *National Geographic*, summed up the studio's philosophy of nature filming at the time. As he reports, the Disney films were to include "no fences, car tracks, buildings, or telephone poles."[17] The company's strict policy in effect ensured that the kind of nature it sought for the films would never be found "naturally." Even the seemingly remote Pribilof Islands featured in the company's first documentary failed to qualify as pristine nature, having been altered by commercial development, particularly by the fur seal industry that grew up around the islands during the late eighteenth century.[18]

Disney's stylized depictions of nature also influenced the kind of stories typically featured in the films. As Alexander Wilson explains, these narratives were never innocent or politically disinterested.

> The Disney movies always told stories, and the stories always began at the beginning—the spring, the dawn, the birth of a bear cub or otter. They ended at the beginning too, with words like new life, rebirth, hope. . . . Yet for all they opened up and "revealed" of life, the early Disney movies also came with their own constricting logic. The animal stories they trafficked in were among other things transparent allegories of progress, paeans to the official cult of exploration, industrial development, and an ever rising standard of living. Those blooming flowers in "living colour"—a signature of Disney's film work—legitimized our metaphors about economic growth. The flowers were typically shown only to the point of "perfection." Rarely did we see them fading, decaying, consumed by microorganisms that returned them to the earth. . . . Like nineteenth-century accounts of the "winning" of the American West, these postwar nature stories were told over and again. They were fictions of victory for the new Century of Progress.[19]

Disney's ecologies, in effect, played into larger U.S. fantasies about a postwar American way of life; arranged and ordered in socially acceptable ways, the "wild" world that was featured in the films was always an idealized nature presented in a manner that appeased white middle-class audiences. And while the places the Disney filmmakers encountered might not live up to popular expectations, these sites could always be altered to fit industry standards through new technologies in filming and editing.

Critic Richard Schickel tells a story about how Walt Disney himself once telegrammed nature photographer Al Milotte after viewing the rushes of an unreleased documentary on postwar Alaska. "Too many mines. Too many roads. More animals. More Eskimos."[20] Although Alaska was hailed as a Last Frontier during the 1950s, the postwar period also experienced tremendous changes that complicated the studio's attempts to figure it as a balanced, pristine nature. The war effort in Alaska, for instance, led to the building of a vast military industrial complex in the region that included several army bases and a highway system that better linked Alaska with the rest of the United States. During the war years, however, Alaska's position at the margins of the United States often evoked anxious national responses. While the 1867 purchase of Alaska from Russia offered the United States a new staging ground for territorial expansion, that same act now placed the country in dangerous proximity to the Soviet Union. The incorporation of Alaska that had dramatically transformed the nation's borders in the nineteenth century created a gap between the region and the rest of the country, which made Alaska particularly vulnerable to possible attack. After the Japanese invasion of the Aleutians in 1942, Alaska became an even more unsettling region in the dominant American imagination.

Anxieties about the nation's military and environmental policies, however, were effectively reworked and reframed in Disney's True-Life Adventure series as the nature themes featured in the films overshadowed the military realities shaping the Far North. The Disney nature documentaries in the 1950s and 1960s clearly supported the protection of wild land, often depicting their settings as intimate, inviting spaces while erasing signs that these regions had already been altered by the nation's military-industrial complex. At the same time they focused attention on "natural" spaces, Disney films in turn domesticated the wild, bringing once remote terrain into the close realm of the viewer but removing any social element that might disturb the aesthetic requirements of the films. As a result, critics over the years have commonly faulted the nature documentaries for creating high expectations in audiences who often seek the same nature upon venturing into the wild, only to be frustrated by what they actually experience.[21]

Crisler herself began to recognize that even Alaska failed to provide an adequate sanctuary from the nation's Cold War realities, and she interrupted her narrative at various points to criticize the changes overcoming the North. According to her, the Distant Early Warning radar system, or DEW Line (which was rendered obsolete upon completion), represented a particular threat to Alaska as a wilderness sanctuary. Designed to detect a possible Soviet missile attack, the development of the DEW Line had already profoundly disturbed the tranquility of the region and promised to keep "wiping out wild habitat and animals in the biggest, best-armed invasion of this fragile life zone . . . ever performed." Crisler also feared that the military infrastructure in Alaska would make the region vulnerable to a future invasion of another sort. Commenting on the military presence in Alaska, she complained that "[h]ardly a man of the thousands to come, on hundreds of planes and ships, but would want to kill. Kill a polar bear, a walrus, foxes, a wolf—everything in fact that moved"

(*Arctic*, 300). Her concerns were perhaps well founded; by 1952, just over half of Alaska's workforce was employed by the Department of Defense, whose military bases were located primarily near Anchorage and Fairbanks. The region's expanding population and economic growth, both of which were fueled by the war, also changed the ethnic makeup of Alaska. Prior to World War II, the region's population was roughly 50 percent Alaska Native and 50 percent Caucasian; by 1950, however, the population shifted to 25 percent native and 75 percent nonnative. The war brought national attention to Alaska as never before; and the settlement of white Americans was eased by the built environment the Cold War helped bring to the region. Crisler thus witnessed a military buildup that promised to change the Alaska she admired in both profound and unimagined ways.[22]

If Crisler expressed frustration about the new militarized landscapes in Alaska, she would soon be vexed by the challenges the region's natural terrain posed as well. She and her husband initially planned to film the migration of caribou herds across the Alaskan Arctic, but they had difficulties capturing enough footage during their stay in the region. When bad weather hit, the two were forced inside their small canvas tent for long periods of time. In her narrative, Crisler expressed anxiety about the delays they faced; as she explained, their window of opportunity for filming was particularly tight, as the best time of year for wildlife sightings came in the spring and fall during the birthing and mating seasons. After a disastrous start, Crisler began to recognize their folly, confessing that the two of them had approached their project with "unconscious arrogance," thinking of the Arctic merely as the field for their work (*Arctic*, 15). Overly confident that they not only could survive the challenges of life in arctic Alaska but also capture the required shots for Disney, she was dismayed by the difficulties that hindered their project. Although she was at first reluctant to use captive animals to complete the footage, Crisler eventually began to recognize the benefits of doing so; her husband's seventy-five-pound camera and all their supplies could now be managed with greater ease, and scenes of "everyday" wolf life could be obtained in greater detail. The capture of the wolves, of course, remained off-screen, an aspect of filmmaking concealed from the audiences of *White Wilderness*. Just as Disney himself sought to hide the mines, the roads, and all other signs of modern life encroaching on Alaska while focusing on "animals" and "Eskimos," his filmmakers also frequently erased the conditions that enabled their own filmic productions. Alexander Wilson argues that such detachment was always a crucial illusion promoted in the nature films and that, as a result, many of them "don't reveal the deep involvement with nature necessary to their making: large crews, helicopters, camera blinds, sets, telescopic lenses, remote sound, and trained animals flown in from another part of the continent."[23]

The Crislers apparently were not the only filmmakers involved with White Wilderness who actively staged nature. According to reports, some cameramen working on the documentary were asked not only to film a sequence involving the migration of lemmings but, if necessary, to throw them over the seaside cliff to create the image of thousands of lemmings engaged in mass suicide.[24] The scene in turn has become one of the most widely referenced moments in the True-Life Adventure Series. Likewise, while it was a commonplace to employ captive animals in filming wildlife, it was also a widespread practice to train them to respond a certain way in front of the camera, or, if they failed to do so, to forcefully encourage them to act in the required manner. According to Wilson, this method was so central to the production of nature documentaries that in the Disney wildlife films, an animal's performance in front of a camera was typically presented as actual animal behavior.[25]

Over the years, these production practices have been criticized by scholars who bemoan what they call "tabloid naturalism" or the "Disney-ification" of the wild.[26] In the nature films, the wild world is often turned into domesticated space, docile and subdued. In the process, the world outside our homes and cities is transformed into sites for consumption through the studio's use of new photographic technologies. As Wilson remarks, "[t]he camera, with its insistence on perspective and the narrow field, exaggerates the eye's tendency to fragment, objectify, and estrange. Staring through a viewfinder, we experience the physical world as landscape, background. . . . [T]he snapshot transforms the resistant aspect of nature into something familiar and intimate, something we can hold in our hands and memories." These photographic technologies, in turn, allow viewers to assume control over the "visual environments of our culture."[27] Karla Armbruster puts it another way, explaining that wildlife photography establishes dominion over nature by creating in us "fixed, dualistic relationships with the natural: viewer and viewed, self and other, subject and object."[28]

For Crisler, the problems of managing nature ended up shaping her experiences in Alaska in several ways. After her husband arranged for the delivery of a litter of wolf pups, she discovered herself in the awkward position of raising captive animals in the wild. Over the next several months, however, Crisler grew increasingly fond of these "four-legged" creatures, whom she named Trigger and Lady, and eventually allowed her husband to arrange for five more captive pups to join their wolf family. Once their filmmaking mission for Disney ended and they prepared to leave Alaska, the Crislers were left pondering the fate of the animals they had kept in captivity over the course of several months. Lois Crisler realized that the wolves she had raised during this time could not in any case survive by themselves outside captivity, as they were now completely dependent on her. The problem was nothing less than a life-and-death matter; if she let the young wolves run free in the arctic wilds, the animals would be sure to perish, unable to feed or care for themselves in a land whose challenges and difficulties they had not yet mastered.

Having domesticated an aspect of the wild world, Crisler found herself cast in a strange role as a kind of modern-day Frankenstein, the scientist who alters relations between nature and culture, creating an Other both horribly powerful and fully dependent on its maker. The narrative shaped her experiences with wolves on various levels; like Mary Shelley's scientist, Crisler frequently had to ward off physical attacks from her "creations," and at one point, went to great lengths to placate them by finding a mate for one of the wolves. At the same time, she found herself both employing and rejecting aspects of a script that has long framed white women's experiences in the wild, a script found in the captivity narrative. This form of writing addressed the experiences of European settlers taken captive by Native Americans. Although some men utilized the genre, it belonged largely to the realm of white women writers. The most significant captivity narrative to appear in North America was written in 1682 by Mary Rowlandson, a Puritan. The account, which told of her trials and eventual restoration, established the conventions of the form. Utilizing Christian allegory to explain these events, the narrative indicated that the captor's restoration involved a spiritual as well as a social dimension. As a national genre that shaped meanings about multicultural encounters, the captivity narrative has operated as a kind of grammar for white women's wilderness experiences. Historically, the captivity narrative has worked to convince the settler community of its moral standing while also providing a kind of innocence for the white female settler in the New World. As Michelle Burnham points out, these narratives always seek to erase the history of border transgressions, territorial dispossession, and colonial violence. By insisting on national innocence and moral rightness, captivity narratives became centrally positioned within a rhetoric of American exceptionalism.[29]

At times, Crisler relied on this strategy of innocence in describing her experiences in this "new world." As an active agent who helped arrange captivity for the wolves, however, Crisler also found she could not fully lay claim to this tradition and even confessed that she felt herself serving as "jailor" to these once wild animals.[30] No longer the innocent party in her encounter with the wild, Crisler discovered herself in a new subject position, this time as the guilty captor. While she continually reproached herself about her role in domesticating wolves, Crisler's experiences encourage us to reconsider the complexities shaping white women settlers and their encounters with new landscapes. Like the captivity narrative, which plays into a history of exceptionalism and American national innocence, the theories that abound concerning women and nature also frequently rely on a logic of female exceptionalism.[31] In ecofeminism, that body of theory that more than any other addresses questions of gender, culture, and nature, women are often uniquely positioned to address the concerns of the environment. Ecofeminists typically recognize that while women as a group are not essentially closer to nature, they, like racial or class minorities, often have been placed in the position of the Other, treated as nature in the realm of male culture. Greta Gaard, for instance, defines the authority of ecofeminism, explaining that "the ideology which authorizes oppressions such as those based on race, class, gender, sexuality, physical abilities, and species is the same ideology which sanctions the oppression of nature." Ecofeminism, in turn, responds by positing "a sense of self most commonly expressed by women and various other nondominant groups—a self that is interconnected with all life."[32]

A problem arises, however, when we consider the complexities of ideology and its specific operations in culture, and when we recognize that we are not always dealing with "the same ideology" or with the same "woman." As Donna Przybylowicz reminds us, subordinated groups have specific histories. Otherness is therefore always contingent and contextual, never fixed or static. "This 'putting into discourse' of general 'Otherness' is an ahistorical and universalizing gesture," Przybylowicz contends. "[It] avoids the necessary consideration of the diversity of oppressed groups, of the political and socio-economic factors involved in the whole question of marginality."[33] Insofar as they posit a monolithic Other as well as a universal Woman, ecofeminists often fail to recognize the authority and power some women have over other groups of people and over the natural world as well. Carolyn Merchant has addressed this problem, explaining that there is "no simple relation between the ways in which nature has been gendered both positively and negatively as female over the past two and a half millennia and the roles of women in society."[34] Merchant's insistence that we question the woman/nature connection to see how it has been played out historically and culturally is important for any inquiry that seeks to unlink women from nature and that tries to reconfigure the gender/culture/power nexus as it operates in specific sites. In this way, Crisler's own experiences in Alaska highlight the problems that arise in simply casting all women as somehow in close allegiance with the natural world.

What is interesting about Crisler's account of filming Alaskan wildlife is that it also expresses a preoccupation with control over nature, precisely the issue ecofeminists criticize about male culture. After confronting the difficulties of Arctic living, for instance, she confessed that "[l]ife outdoors is not one of idleness" (*Arctic*, 8). As she explained, "[t]wo people set down in the wilderness with their possessions are confronted not by dreamy leisure but by hard work" (80). The process of filming as well as the work required to make a home for themselves in the Arctic were much greater challenges than either one of the Crislers had initially anticipated. For a year and a half, the two of them had to coordinate their lives around the schedule of a bush pilot who dropped off food and supplies for them, as well as news from home. While they greeted fresh food and other supplies with pleasure, they hoarded the letters they received from faraway friends, taking care to read only one letter a day in order to spread out their joy over the weeks. Like the pioneers that John McPhee encountered in Alaska during the 1970s who tried to leave the world behind only to become fully dependent on the supplies they managed to fly in, the Crislers were often dismayed by the degree to which they relied on the very culture they thought they were leaving behind. In a telling way, Crisler confessed early in her narrative that "[l]iving off the country in this age of the world is an anachronism except in a real, unavoidable emergency" (28).

Facing these concerns, Crisler often relied on domestic practices which enabled her to assert authority over the terrain she inhabited. At times, her account reads like a woman's conduct book or how-to manual for establishing "homely order" in the wild (14). At the beginning of the text, for instance, Crisler describes in detail what they packed for their excursion as if to establish guidelines for future travelers about the gear they'd need in the Far North. Her text also provides a high-protein recipe for using powdered eggs in hotcakes, an account of their home's construction, including the materials they used to build it, and details about gardening in the Arctic (17, 110, 246). Particularly proud of their crop of radishes, she claims that theirs might be the first ever grown outdoors north of the Brooks Range (246). Crisler's domesticating gaze also extended to the world outside her arctic home. Addressing the living conditions of the region's wildlife, for instance, she expressed pleasure in noting that the tundra, "a thin moss-and-lichen blanket spread over fathomless ice," functioned as "carpet and table" (43). Devising ways of making the North a familiar space was a necessary task for her, as the region frequently seemed to be an inscrutable, perplexing terrain. At one point, Crisler claimed that the Arctic was "unlike anything else we had ever seen," appearing as a "white dimness," a place of "uncanny glory and strangeness" (22, 52).

Recognizing the ways domestic rhetoric often serves larger nationalist practices, critics have begun rethinking the assertion that white women writers, compared to their male counterparts, have traditionally produced less intrusive "imaginative constructs" in accommodating themselves to the difficulties of western settlement.[35] Understanding domesticity as an ideology, feminist scholars have taken a new look at the tropes white women employ in becoming authoritative figures in the literature of travel and exploration. What once seemed a welcome alternative to male conquest for an earlier generation of feminist scholars is now being subjected to new critical scrutiny.[36] In fact, because white women's authority in nation-building projects is often linked to their association with the domestic sphere, critics like Anne McClintock argue that we need to recognize domesticity not only as a site, but also as a social relation to power, a social relation that is not necessarily innocent or opposed to territorial domination.[37]

Reexamining the complexities of white women's social power, Amy Kaplan likewise traces what she calls the "imperial reach of domestic discourse."[38] As she explains, the domestic typically functions as a complex term that refers not only to the space that is in opposition to the public, but also to the space that is distinguished from the foreign. "When we contrast the domestic sphere with the market or political realm, men and women inhabit a divided social terrain, but when we oppose the domestic to the foreign, men and women become national allies against the 'alien' and the determining division is not gender but racial demarcations of otherness."[39] According to Kaplan, domestic rhetoric not only unites men and women within a national context but also helps produce notions of the "foreign" against which the nation can be imagined as home. If domestic discourses serve a crucial role in designating home and Other, then women, situated as caretakers of domestic space, often play a central role in defining the boundaries of the nation.

Just as feminist critics have shown the ways domestic and public spaces are shifting entities that often overlap, Kaplan is also interested in foregrounding the ways the domestic and the foreign operate as fluid entities that likewise come into being in intimate relation with the other.[40] Domestic rhetoric helps produce home as a "bounded" interior space that is somehow opposed to the "boundless" exterior space of the foreign. During times when nations expand, however, domestic discourses frequently serve an important function by mediating the uncertainties that arise in incorporating new terrain. In this manner, white women are often complexly situated in nation-building projects. Domesticity, according to Kaplan, often "entails conquering and taming the wild, the natural, and the alien." In this sense, domesticity is "related to the imperial project of civilizing" and serves as a marker that distinguishes civilization from savagery. "Through the process of domestication, the home contains within itself those wild or foreign elements that must be tamed; domesticity not only monitors the border between the civilized and the savage but also regulates traces of the savage within itself."[41] Kaplan's argument leaves little room for critics to imagine that white women's domestic rhetoric and domestic practices are somehow more benign than the rhetoric and practices of white men; any notion of domesticity as necessarily outside the operations of domination suddenly becomes a difficult idea to uphold.

In Crisler's account of Alaska, home operated as both domestic and wild space, functioning precisely in the dual manner that Kaplan has described. While the domestic sphere Crisler and her husband created for themselves in the Arctic may have provided a sense of security for the two of them, it was barely capable of keeping out the elements. The wolves that became domesticated pets and provided companionship for the Crislers also turned out to be wholly unpredicable and prone to wild behavior at any given moment. Likewise, Alaska itself functioned as both domestic and foreign terrain; at the time Crisler traveled to the Far North, for instance, Alaska was U.S. territory but had not yet been granted statehood. As debates about statehood raged throughout the country during the 1950s, the region's status as an American terrain or domestic site remained unstable. Crisler's task thus involved managing the borders between the wild and the tame as well as the borders between the foreign and the domestic. Such boundary work functioned on various levels in her texts—on the level of nature as Crisler managed the tensions between domesticity and the wilderness, particularly in terms of the animals she and her husband kept in captivity—and on the level of multicultural contacts as she negotiated the problems that arose in confronting her racial Others, the region's original native inhabitants. Domestic rhetoric for Crisler thus operated in complicated ways, serving as a means of managing both the natural and social concerns that shaped the margins of her home space.

At one point, Crisler described the difficulties of supervising what appeared to be an uncontrollable environment, the difficulties of containing what seemed to be a fully wild nature. "You get tired of smallness and hard work. You crave power as you crave a drink of water. Then some exultation of well-being and wild power around you, the great power of wilderness, catches you; you break into cosmic gaiety, pretend you control the uncontrollable. You defy the Great Powers and don't even need words to do so" (*Arctic*, 54). Crisler's descriptions of this wild but compelling nature call forth the tradition of the sublime, a landscape convention that has a long and complex history. Kate Soper addresses the sentiments aroused by the sublime in Western culture. As she explains, while it typically fosters a sense of human smallness in the midst of an overwhelming nature, a sense of human insignificance in the face of divine creation, the sublime has also called forth other feelings. According to Soper, the sublime is at once "appalling because we cannot accommodate the immensity with which it confronts us, and wonderful because this failure itself indicates the superior power of human reason to anything encountered in the natural world."[42] For the adventurer who faces the sublime and manages to thrive in spite of it—the mountaineer, for instance, who bags the treacherous peak or the wilderness enthusiast who pits herself against the arctic wilds—the sublime may also function as a means of establishing human dominion over nonhuman nature. In this sense, the sublime may signal not only fear of the wild but also the achievements of a culture now "in a position to court, rather than simply take flight from, the terrors of nature."[43]

Crisler's own sense of the wild was clearly marked by these two sentiments; at times she felt overwhelmed and insignificant in the face of nature, while in other moments she relished the challenge of carving out a living space for herself in the Arctic. For Crisler, domesticity helped negotiate these ambiguous feelings, as home became a stable center in the midst of a terrifying landscape and domestic metaphors a way of gaining control over her natural surroundings. At times when her power became challenged or thwarted, Crisler employed another rhetorical logic to address her situation. During a particularly heavy storm, for instance, she complained about the snow, which "drove by from endless arctic wastes . . . in a ground blizzard that punished the eyes, burned an exposed wrist like fire and built hummocks under the tent floor." The snow eventually invaded their tent, covering everything from their camera equipment to their bedding (*Arctic*, 18). In their ignorance, the Crislers had neglected to bring a shovel or saw, which would have been useful for cutting snow blocks to make windbreaks or temporary shelters. Realizing the degree to which they were ill-equipped to handle the challenges posed by arctic living, Crisler resorted to a military rhetoric, complaining at one moment that they had "no second line of defense" against the weather (21). Playing with this language, Crisler later referred to the building of a more stable home as "Project Crackerbox" and talked about their departure from Alaska as "Operation Heartbreak" (110, 295).

In the nature film *White Wilderness*, a similar language shapes descriptions of arctic wildlife as the voice-over narration frames nature and animal behavior in a militarized language. During a sequence featuring spring break-up in the Arctic, for instance, the rivers are described as winning their "battle" against the ice as running water breaks through the frozen confines. Later a polar bear cub sets up his brother as a target for a snowball. The narrator describes a "near miss," and then the camera follows another attempt with the voice-over providing commentary—"ammunition ready . . . aim . . . BULL'S EYE." Although the voice-over narrations used by Disney are notorious for the rather patronizing manner in which they address the nonhuman world, this militarized rhetoric in *White Wilderness* is also noteworthy for the ways it signals the return of the repressed: even as the studio sought to present the region as a remarkable Last Frontier, the built environment in Alaska and the military realities of the Far North nevertheless managed to assert their presence. Such language reminds us that Alaska's status as the nation's last frontier also served other important functions in this era. In the Cold War period, as Americans faced a nuclear future that appeared increasingly more frightening each day, encounters with wild nature—places seemingly devoid of the military threat—provided a sense of solace and comfort unavailable to them elsewhere.[44] Wilderness ideology in particular promised to alleviate anxieties in an era profoundly altered by the nation's emerging technologies. Yet as the Crislers themselves discovered while filming the nature documentary for Disney, the military realities of the period were often impossible to escape; even a place as seemingly remote and isolated as the Alaskan Arctic was already threatened by a Cold War military infrastructure.

Anxieties about national security and the complexities of human relations with nonhuman nature became the basis of a popular Hollywood film about Alaska released a few years before the Crislers traveled to the North. In Christian Nyby's 1951 science-fiction movie *The Thing*, Alaska figures as a neglected but important national terrain. The film imagines the post–World War II threat as an alien creature, both a Soviet and a nuclear menace. The Thing, for instance, literally falls out of the sky, contaminating the land and threatening to destroy American civilization as we know it. The film's opening provides audiences with a shot of the Arctic as a barren, wind-blown landscape and later cuts to a scene in an officer's club in Anchorage. The interior of the building is featured as a cozy, warm, and comfortable space which contrasts with the frozen, forbidding outdoors. The officers in the club occupy themselves with card games while fantasizing about the South Sea islands they'd like to retire to in their golden years. Their playful banter breaks up, however, when they receive word that Dr. Carrington, an American scientist at the North Pole, needs their help. Carrington, we learn, is "the fellow at Bikini," an atomic scientist and winner of the Nobel Prize. When he sends a message that an unusual aircraft has crashed near the vicinity of the science station, the officers immediately suspect it to be a Soviet aircraft because, as one character says, the Russians have been "all over the Pole like flies."

After conducting tests on samples recovered from the Thing, the officers discover that the national threat at the Pole is composed of vegetable matter. Appearing as "some form of super-carrot," the alien has managed to construct an aircraft capable of flying millions of miles through space "propelled by some force we don't understand." One of the scientists suggests that the carrot might have come from a planet that allowed it to develop a mind. A reporter at the scene is not convinced: "You mean there are vegetables on this planet that can think?" he asks. The scientist sets him straight, explaining that vegetables on earth have the same potential. "Telegraph vines have intelligence systems" that they use to send messages to other vines, he explains. "Plant intelligence is an old story, even older than the animal arrogance that has overlooked them." Fearing that the Thing might try to colonize Alaska, the officers ponder how to destroy it but are stumped trying to figure out what "you do with a vegetable." Fortunately, the film's only female character is on hand to provide domestic counsel and suggests that the men try to boil, stew, bake, or fry it. The officers take her advice, opting to use kerosene, fence wire, and electricity to eradicate the threat, which is now figured as a danger not only to national security and military development, but to the nation's environmental health as well.

Just as the characters in the film are not protected by the remoteness of arctic Alaska, the Crislers also began to recognize that the region would not provide an adequate sanctuary for the nation's Cold War realities. Lois Crisler eventually addressed this problem with her own version of domestic containment, burrowing into a private, seemingly well-bounded space—her home in the wild. Historian Elaine Tyler May describes the ways U.S. domestic ideology offered a buffer in the postwar period as the domestic version of containment figured home to be the nation's best defense against foreign threats.[45] Yet even as Crisler tried to create a safe zone for herself in the Alaskan landscape, an arctic home complete with a garden, some pets, and a warm kitchen, her attempts to find sanctuary in both the domestic realm and the Far North were not enough to keep at bay the threatening elements of the Cold War.

The tensions she encountered between domesticity and the wild, between culture and nature, for instance, also framed her experiences with the captive wolves used in the footage for *White Wilderness*. As one of the most maligned predators in Western history, wolves have often functioned as symbols for the dangers of the wild. As recently as the 1990s, during Governor Walter Hickel's administration, for instance, wildlife management polices in Alaska still called for widespread wolf eradication across the state.[46] The Crislers' decision to film arctic wolves several decades earlier was thus part of a larger counter-movement among nature enthusiasts to force a reconsideration of national wolf policy. In doing so, however, nature advocates also unwittingly encouraged a reconsideration of definitions of human identity. John Berger describes the ways animals have been used symbolically in the modern era to mediate identity and delineate the boundaries of the human.[47] For Berger, the modern is marked not only by a reinvention of the animal as symbol but also by the gradual disappearance of animals from everyday life. One aspect of the modern, he argues, is the idea that humans appear to live outside or beyond the domain of wild animals.[48] Physically and culturally marginalized from everyday life, animals in turn become positioned in anachronistic space; placed in a "receding past," they are treated as symbols of innocence, as signs for what we have lost with the advent of modern culture.[49] Precisely at the moment that they recede from the familiar, however, captive animals begin to take their place. In the modern world, animals thus reappear in homes as pets where they are co-opted as members of the family, and in zoos or nature preserves as signs of a nation's expansionist practices, symbols of "the conquest of all distant and exotic lands" now brought home to the domestic realm.[50]

In *Arctic Wild*, Crisler describes the thrill of coming into contact with wild animals in Alaska. As many critics have argued, human/animal encounters often have been shaped by larger cultural concerns. Theodore Catton argues, for instance, that from the late nineteenth century onward, wildlife played an important symbolic role in establishing Alaska's status as anachronistic space, helping to designate the region as a primitive, archaic place. Once Alaska became mythologized in the national imagination as a unique terrain, wild animals became central to its image as a Last Frontier. While figures such as John Muir and Jack London did much to shape national responses to the North, the early-twentieth-century sportsmen writers, especially those who focused on big-game hunting, also helped position wild animals as central to the region's frontier mythology.[51] This image held enough power in the national imagination that by the mid-twentieth century, Crisler could argue, "Wilderness without its animals is dead—dead scenery." According to her, the rare person who encounters an "other-than-human being" in close proximity in the wild has the good fortune of becoming a kind of "Cortez," someone who can claim an "authentic delight of spying new worlds." Yet if wilderness without animals operated as dead scenery, Crisler would also go on to argue that animals without wilderness "are a closed book." She and her husband decided to film animals in their natural habitat, Crisler explains, because the two of them believed that "the last frail wonderful webs of wilderness now vanishing from Earth are of some infinite value, a value only sensed and very deep—and liable to perish and be lost in this day" (92, 158).

In their mission to collect images of wild nature before it was forever destroyed, the Crislers ended up violating a larger sense of the wild by forcing it into captivity. If it has been a mark of colonizing nations to bring home specimen and trophies of their adventures, wildlife photography itself may be considered part of this larger project, becoming a means of collecting faraway sights and scenes of exotic nature for American audiences back home. The Disney mission for the Crislers thus functioned much like nature writing did for John Muir and Samuel Hall Young during their trips through the Inside Passage, enabling them to collect and domesticate the wild, exerting control over a foreign element in the process. Crisler explains this problem clearly, telling of their disappointment in missing a "nugget," or the opportunity to capture a special wildlife moment on film. Crisler's husband, Herb, would elaborate: "Getting an Alaska picture is like the gold rush. The stuff is there," he contends. "You spend your spirits, your life and your fortune. Maybe you get it and maybe you don't" (34, 88).[52]

Although Lois Crisler believed that a region required animal inhabitants in order to function as a true wilderness, the dispossession and displacement of the land's human inhabitants became a requirement for the kind of wilderness adventures she hoped to experience. For instance, her relationship with her nearest neighbors, the Iñupiat Eskimos of Anaktuvuk Pass, became complicated precisely as she sought to secure domestic authority for herself in the region. As feminist critics have pointed out, a central aspect of domesticity has often been racial management; domesticity frequently aids colonialism by turning natives into foreigners in their own country.[53] Crisler's domestic ecologies in Alaska worked in this capacity as she lamented the Eskimos' practices toward animals, implicitly indicating that their treatment of wildlife marked them as uncivilized and poorly domesticated. Worried that the Eskimos might shoot their captive wolves in order to collect a $50 bounty from wildlife authorities in the region,

Crisler took special care to ensure her animals were always well secured. When two Eskimo brothers, Jack and Jonas, hunted in close proximity to their arctic home, the Crislers tried to strike an agreement with them that they would kill no animals but caribou within their camera range (208). Herb Crisler later criticized the Eskimos for their hunting practices. "Like throwing stones into the herd," he remarked. "They don't try to hit any particular one" (204).

Lois Crisler likewise tried to explain the Eskimos' animal practices as stemming from their position in anachronistic space and, in doing so, employed a common trope among white settler communities who often depict the native inhabitants of a region as poor stewards of the land.[54] As she suggested, Jack and Jonas were members of "one of the last two groups of nomads left in North America" and thus had the misfortune of only recently leaving the "Stone Age" (200). Crisler went on to elaborate: "The Eskimos did not bother to follow and kill the wounded. . . . It is well to understand that Jack and Jonas killed and felt pride in doing it not from malice but from historic background. Eskimos did not kill unduly, were not out of balance with their environment, when white men first came. They hunted skillfully but with native weapons. Now they are armed lavishly with white men's weapons but retain the imperative of men armed only with Stone Age weapons—kill regardless" (205). In an interesting reversal of the historical encounter between whites and Eskimos in Alaska, Crisler figures herself as the victim of an invading culture whose less-than-ideal hunting procedures and animal practices threaten the safety of her filmic mission and her captive wolves. In this description, the Eskimos are depicted as somehow encroaching on *her* land and her domesticity, as somehow impoverishing the fragile environment she and her husband needed in order to the complete the assignment for Disney. Donna Haraway describes a similar system operating after World War II among white primatologists in Africa and Asia. According to Haraway, many of the researchers lacked an understanding of their own position within systems of racism and imperialism. "Many sought a 'pure' nature, unspoiled by contact with people; and so they sought untouched species, analogous to the 'natives' once sought by colonial anthropologists. But for the observer of animals, the indigenous peoples of Africa and Asia were a nuisance, a threat to conservation— indeed encroaching 'aliens.' "[55]

Crisler's depiction of the Eskimos as trapped in tradition, confined to anachronistic space, enabled her to position herself as perfectly housed in the Arctic. In order for her to function adequately as a true sojourner to the wild, however, she needed a racial Other that she could position as primitive or archaic. Because a central element of wilderness ideology suggests that the figure who most benefits from wilderness is the figure who is most removed from it, wilderness ideology requires an ecological subject who is able to throw off excessive layers of culture. White women, insofar as they have been associated mostly with nature, are often regarded as being too far removed from culture to function as proper wilderness subjects. Crisler thus required an Other caught in tradition in order for her to claim a space for herself in this wild world. At one point, for instance, she expressed her unease at taking up the role of the modern arctic pioneer and chastised herself for not living up to the role of intrepid adventurer. Meanwhile, she lamented the plight of her husband, who must face the wilderness "with only a woman to help him" (*Arctic*, 210). Anachronistic Eskimos, however, helped alleviate Crisler's dilemma to a certain extent by occupying a point seemingly more removed from "culture" than she.

Yet the racial Other she located in Alaska also occupied this anachronistic space somewhat ambivalently. The Eskimos she encountered, for instance, had already been tainted by Euro-American technologies, which they somehow had not learned to use correctly, according to Crisler. The Eskimos thus improperly fit the role of the pristine Noble Savage even as they seemed improperly civilized. As she noted, they hunted animals with rifles rather than with "native weapons," a move that signaled their fall from harmony and marked their cultural degradation. To a certain extent, then, Crisler seemed to understood the problems in considering Alaska an untouched wilderness space located outside the modern, even if she didn't go so far as to replace this vision with something else or fully understand her role in its political history.

Berger's argument that nonhuman animals often become intimately connected to visions of progress and understandings of human identity is especially compelling in this instance.[56] As other critics have noted, animals frequently function as a site of struggle over the production of cultural differences, serving to mediate in processes of modernization and racialization. Haraway argues, for instance, that animal societies have been used extensively to naturalize "the oppressive orders of domination in the human body politic."[57] This practice is seen as cultures set up divisions concerning who is human and who is animal, as well as boundaries that delineate the proper uses, treatment, and understanding of animals.[58] The act of policing human-animal borders by regulating animal practices serves to maintain certain notions of human identity even as it legitimates the animal practices of the dominant group.[59] Violations of the human-animal boundary thus enable cultures to link what they regard as beastly people with "people-acting-beastly" toward other animals.[60] In Crisler's text, Eskimos become foreigners in their own land, made alien to their environment because their animal practices somehow keep them from entering the realm of the civilized. At the same time, Crisler's animal practices—her capture and domestication of the wolves for use in the wildlife documentary—somehow did not place her in the same position, in a site outside civilized culture.

Crisler's condemnation of the Eskimos' animal practices, however, did not prevent her from questioning some of her own animal practices. After leaving Alaska and settling in Colorado, she tried to create a life for the wolves that replicated the one they had left in the North. Domesticity ultimately failed both her and the wolves; when her marriage with Herb ended, Lois Crisler was forced to sell her cabin and was later left with no place to house the animals. She eventually found homes for some of them but decided to put the last wolf to death rather than placing her in a zoo. Upon the wolf's death, Crisler tried to honor the animal's wildness, imagining her no longer in this world but "free with her fellow wolves on the tundra in the old big days of her youth" (Captive, 237).

Ultimately, the domestic ecologies that informed Crisler's experiences in Alaska end up telling us more about her own ideas of a national landscape than they do about the natural elements of the region. As her account indicates, white women may have had more in common with their white male counterparts in their understandings of the natural world than many critics have liked to acknowledge. Like Muir, London, and other popular male writers of the region, Crisler was deeply invested in notions of Alaska as anachronistic space, a place that called forth the nation's wild past and that enabled the Euro-American adventurer to experience a primitive frontier world. By recognizing white women's complicity with the myth of Alaska, scholars may be encouraged to reexamine those forms of criticism which also tend to celebrate the anachronistic space that a purportedly universal "Woman" experiences and which argue that women as a group are somehow beyond the culture that secures domination over the world, outside a modern sensibility that gives some humans power over nonhuman nature.

White Women and Wilderness Lifestyles

During the years she spent fighting for the preservation of Alaska's wild lands, Margaret Murie gained a reputation as one of the modern environmental movement's most dedicated and steadfast supporters and one of Alaska's most important nature advocates. With her husband, the famous wildlife biologist Olaus Murie, Margaret Murie lobbied for wilderness preservation during the statehood debates in the 1950s and helped ignite an environmental movement across Alaska in the process. She was instrumental in creating the Arctic National Wildlife Refuge in 1960, and four years later, President Lyndon Johnson invited her to the White House after signing the Wilderness Act, for which she also lobbied.[61] Murie served on the governing council of the Wilderness Society and gave numerous talks and lectures across the country about the urgency of preserving the nation's wilderness areas. One of the highlights of her career as a nature advocate came in 1980, when Congress passed the Alaska National Interest Lands Conservation Act (ANILCA), which set aside further public lands for parks and wildlife refuges. During her lifetime, Murie received several prestigious environmental awards, including the Audubon Medal in 1980 and the Sierra Club's John Muir Award in 1981. She likewise played an important role in developing the University of Alaska and was the first woman graduate of what was then called the Agricultural College and School of Mines.[62]

During the years she worked alongside her husband studying wildlife and collecting samples for his biological surveys, Murie also managed to raise three children, who often accompanied her during her many expeditions to the North.[63] Although she would later settle in Wyoming, Murie made several trips to Alaska over the years and continued to seek out wilderness adventures. During one trip she took in 1975 when she was seventy-three years old, for instance, the small plane transporting her crash-landed after its motor failed. Murie and the rest of the crew were forced to wait seventeen hours until a helicopter pilot was finally able to rescue them.[64] While keeping busy with her environmental activism and family life, Murie also managed to write three books, including *Two in the Far North*, an autobiography about her experiences living in Alaska; *Island Between*, a novel about Alaska Eskimo life during the precontact era; and *Wapiti Wilderness*, an environmental account that she cowrote with her husband about their experiences studying caribou populations near their home in Wyoming. With a new edition of her autobiography published by the University of Alaska Press in 1997, Murie is currently being rediscovered by nature advocates, who now consider her a central figure in the modern environmental movement.[65]

Murie's writings provide us with an important record of the environmental changes shaping Alaska and reveal the shifting issues that nature advocates faced in the twentieth century. Her writings are also important for the ways they highlight one of the underlying concerns of the contemporary environmental movement. Throughout her personal writings and in the numerous speeches she delivered to environmental groups, Murie has unfailingly testified to the growing focus on amenity concerns in the post–World War II period. This attention to wilderness lifestyles, which has dominated discussions in the mainstream environmental movement, reveals a strong interest in preserving what many see as a way of living now threatened, a way of living associated with the wild, the frontier, the nation's pioneer past.[66] Having gained important achievements in wilderness advocacy and other quality-of-life issues, however, the mainstream environmental movement has often been cast as a white middle-class organization precisely because of its focus on lifestyle issues. The grassroots environmental justice movement, for instance, has often criticized mainstream nature advocates for failing to speak to the concerns of the nation's economically underprivileged classes and its communities of color, both of which disproportionately face environmental dangers in their living and work spaces.[67] In providing a central voice to the wilderness debates over the last forty years, Murie herself did not escape this criticism. Although she worked passionately to preserve Alaska's wild lands, her environmental efforts tended to address the needs of white middle-class recreationalists, who typically have an interest in preserving the amenity values of a region but who often fail to question how these desires and concerns might impact other cultures.

In a talk she gave in Anchorage in June 1975, for instance, Murie spoke of the changes she noted upon returning to the North after an eight-year absence. The 1970s saw the completion of the TransAlaska pipeline, which was built to carry crude oil over from the North Slope to the port of Valdez in the Gulf of Alaska. The construction of the pipeline boosted Alaska's economy in the 1970s and, in turn, profoundly altered patterns of land tenure in the state. Upon visiting Alaska in the mid-1970s, Murie argued that the region she now encountered seemed both physically and emotionally altered by these developments. Although she felt heartened by the fact that many residents still seemed interested in debating lifestyle issues, Murie also expressed concern that the pipeline might have forever changed what made Alaska unique in the popular American imagination.[68]

Murie remarked that, while other regions had slowly lost their frontier conditions, Alaska still managed to retain its identity as a throwback to the nation's primitive past. Yet even as she celebrated the opportunities available in this Last Frontier, Murie also conceded that actual pioneer living was not always what it was made out to be. Remembering a time when wood was the only fuel available to people, when hillsides were stripped of birch forests to keep the miners warm as they pioneered for gold, and when every household burned ten cords of timber each winter to survive in the subzero temperatures, she recalled a time that was not necessarily concerned with the same ecological issues that shape nature advocacy today. According to Murie, what was most important about the frontier past was that everyone was cared for and everyone counted. As she explained, frontier Alaska was most distinctive for the ways it fostered a spirit of community and belonging.

Such testimony about Alaska's past often formed the basis of her public speeches as well. For more than four decades, Murie gave numerous lectures across the country that popularized the region as a wilderness area and that, in turn, cautioned Americans about the threats facing this Last Frontier. As Murie explained, "there may be people who feel no need for nature. They are fortunate, perhaps. But for those of us who feel otherwise, who feel something is missing unless we can hike across land disturbed only by our footsteps or see creatures roaming freely as they have always done, we are sure there should still be wilderness."[69] In her statement to Congress supporting the establishment of a wilderness preserve in the Brooks Range, Murie again spoke of the "spiritual, mental, and physical" amenities offered by wilderness regions. Echoing the sentiments of Robert Marshall, who also spoke of the importance of preserving the Brooks Range for future Americans, Murie contended that the country and its future generations "will need and crave and benefit from the experience of travel in far places, untouched places, under their own power. For those who are willing to exert themselves for this experience, there is a great gift to be won in places like the Arctic Wildlife Range, a gift to be had nowadays in very few remaining parts of our plundered planet—the gift of . . . personal satisfaction . . . [and] personal well-being."[70]

Lifestyle issues also figure centrally in her autobiography, *Two in the Far North*. In this account, Murie recalls Fairbanks in the early twentieth century, describing a rugged pioneer town that nevertheless worked hard to foster a sense of community. Although it was "surrounded by a wilderness so vast it could not be visualized," the town was populated by people who came together to help one another survive the harsh climate and physical obstacles of the region (28). If life in this Last Frontier challenged the settlers, who struggled merely to survive its difficult conditions, the region also allowed them to cultivate a sense that they were different and set apart from other Americans. For Murie and the pioneer settlers she describes, Alaska was a place where "all things are possible" (151). The region was attractive in large part because it allowed the pioneers to experience a primitive lifestyle; in a world increasingly modernized, few Americans seemed to have the opportunity to live simply. And at a time when the North was still relatively undeveloped by Euro-Americans, these early-twentieth-century settlers believed that America's past could be found in Alaska. For them, the North allowed a return to a more primitive era and provided a space of retreat where the encroachments of modern society had not yet altered the region's frontier possibilities. Quality-of-life issues shaped their decisions to live in Alaska; and throughout her writings, Murie continually addressed the importance of wilderness protection to ensure a pioneer lifestyle, the thrill of living in a world that "seemed to be our very own" (154).

Insofar as it refers to the world of the everyday, to the manner in which individuals and communities create homes for themselves in environments of their own choosing, lifestyle is intimately tied to the concerns and the world of women. If women are frequently granted domain over the domestic sphere, the style of one's living often becomes a topic about which they are able to assert authority. White women nature writers like Murie did, in fact, situate themselves in the wilderness debates of the postwar period, commanding influence within the environmental movement by placing their ideas about lifestyle within the rhetoric of domesticity. In that sense, white women were able to position themselves within larger conversations that helped shape the environmental movement, especially as questions about the amenity values of landscape began to dominate the discourse. White women thus gained a kind of authority in the wild by asserting their knowlege about lifestyle and quality-of-life issues related to the home.

In *Two in the Far North*, Murie takes care to show how white women may adapt themselves to a frontier lifestyle by focusing on what they gain from this way of life. She describes her marriage to Olaus in a summer ceremony that took place on the Yukon River in broad daylight at 2:30 A.M. Her trousseau, containing "not a bit of lace or a ribbon," consisted instead of a tent, a stove, duffel bags, snowshoes, a fur parka, wool undergarments, and flannel pajamas (84). Their honeymoon, she explains, was later spent on a 550-mile dogsled ride to the Brooks Range, where her husband was studying caribou herds for a government survey. In Murie's description, pioneer living enables white women to eschew the social restrictions of femininity; wilderness survival, it seems, requires certain gendered forms of identity to be left aside. Although life in the North provided opportunities that were seemingly unavailable elsewhere in the country, at times the actual experiences of making a home for herself in the wild proved to be an enormous challenge. Murie describes in particular the sense of isolation that backcountry living entailed for her as a white woman. The arcadian lifestyle took its toll on her, and she was frequently overcome with a sense of loneliness and alienation. At one point, Murie addressed the "special kind of feeling" that overcomes a person who unexpectedly meets another settler in the wild. "It must be the same feeling our forebears had when, slipping through the primeval forest, they glimpsed the smoke of another camp fire wafting over the treetops" (155). While wilderness was supposed to operate as a place outside modern culture, Murie discovered that it was often occupied by people who deeply felt the loss of social encounters. As she explained, "Most old sourdoughs felt that no camping place should be passed up, no matter how short the travel, perhaps because they had gone through too many days of forced marches and long, long trails" (156).

At one point, Murie describes an instance that captures the other side of wilderness living. While her husband was off conducting field research with other scientists, she was left alone at their settlement, where she struggled to keep herself occupied. Without reading material and not much to do on her own, Murie took solace in thinking how the many lone trappers and miners in the region must have felt in earlier years.

> All my life I had heard stories about their having nothing to do but read the labels on cans. That second day I read every label in the grub bag. The milk cans were especially nice; they had recipes on them. I looked them over, but they had all come out of the same case, I guess, so I memorized the recipe for creamed clams that was on every one of them. Hard to tell when we should ever see a clam! Our ham was wrapped in a page from a magazine, and I shall forever be wondering whether the wandering son got back to the old homestead and his childhood sweetheart before his poor old mother passed away! (181)

Later, when Olaus returned home late from his fieldwork, he was dismayed to find his wife in her sleeping bag "sobbing wildly," overcome with fear. Although she tried to hide her anxieties lest her husband regret his decision to bring her along, Murie believed that she had "better learn not to worry" or else she might end up becoming a "nervous, nagging, unhappy wife" (182, 184).

While her husband was away collecting samples and taking photographs of brown bears during another research trip, their first child was born and Olaus didn't receive word until his son was eighteen days old. Afterward, Margaret Murie decided not to be left alone with their young child and insisted on accompanying her husband and his colleagues on their next collecting trip. In the campsite with only herself for company, Murie suffered the toll of unbearable solitude once more. Upon their return, the men "came thrashing back triumphantly through the brush," she writes, "[only] to find a completely unreasonable, weeping young woman kneeling in the boat. They were thrilled over the adventure of the good bear specimen for the museum. I was hysterical because I had no nets, no hot water, and worried about my poor baby, wondering why I had come along anyway." The process of making a life in the wilds of Alaska often posed problems for figures like Murie, who were not unencumbered by the details of the everyday and the responsibilities of family. Even as her ability to take pleasure in the same pioneer lifestyle that her male counterparts enjoyed seemed circumscribed, Murie nevertheless extolled the virtues of the region in her autobiography, explaining that the value of the land could be located in "its absolutely untouched character." It was not that the region was spectacular or unique, but that it was pure and therefore "just beautiful." Alaska's value thus stemmed from its position as a wild country not yet marked by the "sign of man or his structures." According to Murie, for this feature alone, Alaska was worth saving (209, 219–20, 340).

The Last Frontier that Murie remembered so fondly in her autobiography faced changes over the years that threatened to alter its position as a wilderness terrain. The intrusion of the modern world became a frequent anxiety shaping her speeches and writing. This threat of diminished space, as many critics have argued, has long shaped the nation's literature, and particularly the literature of the American West, which has always lamented the death of one West or another. And in the spatial imagination of the United States, Alaska— the West's annex—has not escaped this fate.[71] During her frequent visits to Alaska over the years, Murie made these concerns a central part of her wilderness advocacy. At one point, she recalled the fate of the prospectors who were displaced by the big mining companies. She then told of the impact of World War II, its army personnel, vast infrastructures, and the "take-over of thousands of acres of land." Murie later described visiting Fairbanks in the postwar period and feeling as if she were "in a strange place I had not known before." Worried that the "character" of the old way of life might be lost, she wondered if Alaska might have sacrificed "a life-style beyond compare in the whole world." Murie also told of other changes altering Alaska, lamenting the "unsightly appearance of the new strung-out communities along the Highway" and criticizing the "quite nice homes [that] were spoiled by heaps of old truck tires, old oil drums, old car bodies, all sorts of junk." The once "charming" town of Sitka, in particular, seemed to face problems according to her. "The ride in . . . from the Ferry terminal . . . is horrifying," she explained. "All this kind of thing in front of glorious mountains and beautiful lakes and streams and forests, made a jarring contrast. There is such terrific potential for beauty in a beautiful environment."[72]

The Alaska that once provided an "oasis of stillness" and that previously functioned as a "great empty quiet land" now faced all sorts of threats from the forces of development (*Two in the Far North*, 231, 234). Upon her return to Alaska in 1975, for instance, Murie wondered if the "new Alaska" of the oil age would be able to retain the frontier qualities the early pioneers cherished or whether it would go the way of the "other states."[73] What remained desirable about Alaska, Murie contended in her speech, was that it promised to offer "a picture of the past" while providing "a virility, a ruggedness, an individual freedom that is fast disappearing."[74] For her, wild nature was Alaska's greatest gift to the nation, serving both scientific and ecological ends. The Alaskan wilderness was likewise important because it provided much needed "elbow room" for humans; as a place largely "untouched by humans," this wild land offered a space for recreational activities and provided inspiration for the "human spirit."[75]

According to Murie and other nature advocates in the post–World War II period, Alaska was an especially important arena for wilderness battles as the region seemed to offer many Americans their last chance to experience a world yet unmarked by the encroachments of modern civilization. More recently, however, environmental critics have addressed the false hope offered by wilderness. Peter van Wyck, for instance, explains that among mainstream nature advocates, wilderness stands as "the realm of the real, the well-spring of life, and as [a] mirror in which is reflected authentic experiences of the world." Regarded as an empty space through which cultures acknowledge their own transgressive presence, wilderness "is presumed to present a surplus of meaning in relation to dominant modes of expression . . . perhaps this would go some distance to account for why wilderness is so often mystified; its contents are presumed to be far larger than its form."[76] Wilderness in this sense may be regarded as the ultimate fetish, an object given inordinate value and thought to be endowed with special powers, an object that captures both a sense of pleasure and anxiety in the eyes of the beholder.[77]

Along a similar line, Richard White addresses what he regards as an ahistorical strain of wilderness thinking that "imagines a self-regulating nature, fully independent of humans, that should be shielded from use," a form of thinking that, as many critics have noted, contributes to the erasure of indigenous people in the land.[78] Meanwhile, for William Cronon, the problem with understanding wilderness as a pristine, untouched nature is that it promises an illusion "that we can somehow wipe clean the slate of our past and return to the tabula rasa that supposedly existed before we began to leave our marks on the world." Wilderness often functions as a particular form of white flight, an alluring escape from history. According to Cronon, this understanding primarily reflects Euro-American notions of nature. Rather than being an empty place promising renewal or a nature unspoiled by contact with human cultures, he points out, wilderness should be understood as a constructed entity that requires human separation from nature in order for it to serve as a sanctuary from the modern world.[79]

These illusions of separation have been complicating factors for preservationist battles in Alaska after 1960. With the discovery of vast oil deposits at Prudhoe Bay, Alaska's status as a Last Frontier was threatened as it had never been before. The proposed construction of the TransAlaska pipeline through the North upset many environmental activists and faced tremendous obstacles, as much of its route passed through areas inhabited by Alaska Natives. Because the U.S. government had not adopted a reservation system or enacted a policy of removal in Alaska to the extent that it did in the Lower 48, the state's indigenous inhabitants had never been fully dispossessed of their lands and had not extinguished their land rights. Before the oil companies could construct a pipeline, then, native land claims had to be resolved. After much negotiation, the federal government eventually agreed to a $925 million settlement with the Indians and Eskimos of Alaska. The Alaska Native Claims Settlement Act, or ANCSA, divided the claim among different corporations that would be run by natives who would make land selections and invest the settlement money for their stockholders.[80]

Theodore Catton addresses the conflicts that emerged between environmentalist groups and Alaska Natives in the wake of the land settlement in the 1970s. He argues that at first, the discovery of oil on the North Slope helped unite environmentalists and native leaders, who began to see each other as potential allies against the government and the oil industry. As time passed, however, differences emerged between the two groups that often set them at odds. Throughout the history of the region as a U.S. domain, for instance, native residents faced unfair treatment under Alaska's various wildlife management laws. As a result, Eskimos and Indians often transferred their mistrust of officials onto wilderness advocates. In addition, many native leaders had to contend with white environmentalists who operated with their own, culturally specific ideas about nature and who believed that the lands the Alaska Natives received would be preserved in a particular manner that fostered a pioneer or wilderness lifestyle as Euro-Americans knew it.[81] The clash between Alaska Natives and white environmentalists foregrounds the different historical experiences of nature that each group brought with them and indicates the ways concepts such as the "wilderness," "frontier," or "quality of life" do not speak to a universal audience but operate as culturally specific notions.

In the 1950s, during the time Margaret and Olaus Murie were lobbying for the creation of a wilderness preserve in Alaska, nature writer Sally Carrighar, who was living in the North, provided a voice of dissent in the nature debates. Concerned about how such a wilderness preserve might be established and maintained, Carrighar corresponded with officials at the National Park Service in 1954, expressing caution about the Eskimos whose needs she feared might be overlooked. Although she supported wilderness preservation as a goal in general, Carrighar explained that she felt wilderness should also be free of official regulation and expressed concerns that government officials might be too eager to fend off the wild from human use. She likewise argued that the designation and management of wilderness areas should be the responsibility of local populations rather than officials who lived far from the actual sites chosen to be preserved.[82]

Her concerns were not misplaced; generally speaking, while wilderness advocates were preoccupied about the death of the wild and the transformation of Alaska, they often didn't concern themselves with the fate of the region's indigenous populations who were being displaced from that world. The lifestyles they were most concerned with were those of white settlers. As Catton argues, the history of land tenure in Alaska ultimately required that the concept of wilderness faced redefinition, as government officials were forced to rethink the idea of wilderness as a nature separate from humans. Although it has been a commonplace for whites to consider themselves "alone in nature" even when they were in the immediate company of a region's native inhabitants, this illusion became difficult to uphold in Alaska.[83] In the national and state laws creating parks and preserves in Alaska, native peoples mostly retained subsistence rights in the region, an act that challenged white American ideas of wilderness as an empty space devoid of human inhabitants. Catton argues, however, that this redefinition did not happen without conflict; although the creation of most wilderness parks and preserves usually allowed the inclusion of native populations, the white residents and wildlife officials in the state were still conflicted about how exactly they were going to accommodate native groups in the region. Alaska thus operated as an "inhabited wilderness" more in theory than in practice.[84]

Insofar as Murie lobbied for the protection of wilderness spaces and wilderness lifestyles, she ended up speaking a particular racial ecology that at times had unintended consequences for native peoples, whose cultural practices and histories were not included in this larger vision. Like Lois Crisler, the frontier experiences Margaret Murie treasured were products of a larger white conception of community and nature. Although Murie and other white women in the post–World War II period were able to create political positions for themselves in the wilderness debates by presenting quality-of-life issues as topics about which they had experience and authority, their environmental visions were culturally specific ones and did not necessarily speak to the concerns of other populations in Alaska.

Racial Ecologies and the Environmental Movement

Some ecofeminists have cautioned critics about developing theories that place women in central positions of authority over the natural environment. Carolyn Merchant warns scholars, for instance, about the dangers of using the supposed connection between women and nature to sanction a place for women in the environmental movement. "Does not such a connection essentialize women as planetary caretakers and green cleaners?" she asks. "Does it not keep women in their place as caretakers of the earth's household?"[85] Just as Merchant cautions scholars about building a movement on essentialized notions of woman, other critics point to the problems of universalizing the experiences of some women. After all, not all women have been regarded as appropriate green housekeepers in the history of U.S. environmentalism. White women like Lois Crisler and Margaret Murie managed to create authority for themselves by using the domestic discourses that positioned them as powerful in other domains, yet their domestic authority was enabled in large part by their status as white women. Crisler and Murie may have had to find new ways of advocating for the wild that differed from what their male counterparts were doing; as white women, however, they also were afforded certain kinds of power over populations through the racial and gendered ecologies they produced. Anne McClintock has addressed the ways white women have often found themselves in positions of "decided—if borrowed—power." She argues that white women should not be regarded as mere onlookers of history but should be recognized as often "ambiguously complicit" in national discourses in their capacity as both privileged and restricted players.[86] Even though women as a group often have been aligned with nature in Western history, Crisler and Murie didn't necessarily feel themselves to be a part of nature, particularly the nature they associated with wilderness. They instead felt strangely positioned in this space and were thus forced to revise and rework the masculinist visions of Alaska they had received.

In a similar way, white male writers found they had to negotiate the visions that they, too, had previously inherited. During the early years of its status as a U.S. territory, Alaska was considered by many Euro-Americans to be a frozen, barren wasteland of little use to the rest of the nation. The writings of John Muir, Robert Marshall, Jack London, Rex Beach, James Oliver Curwood, and others, however, did much to alter popular perceptions of the region, shifting national understandings of Alaska from a useless terrain to a Last Frontier. White women played a central role in this project as well. Although they struggled to position themselves as authorities in the region, they often did so by framing their experiences within a domestic rhetoric that relied on certain racial privileges. What we learn, then, is that there is no simple dichotomy between the visions and understandings that white men and white women brought with them to Alaska. While white women nature writers may have used a different rhetoric, they often aided the same larger project of using conceptions of nature to draw Alaska more fully into the national orbit.

CHAPTER FOUR

BEYOND THE WHITENESS
OF WILDERNESS

Alaska Native Writers and
Environmental Sovereignty

The last 150 years have seen a great holocaust. There have
been more species lost in the past 150 years than since the Ice
Age. During the same time, Indigenous peoples have been
disappearing from the face of the earth. . . . There is a direct
relationship between the loss of cultural diversity and the loss
of biodiversity.—Winona LaDuke, *All Our Relations*, 1999

The beauty of a mountain means nothing if there is no
passionate commitment to the human society dwelling
in its shadow.—Mary TallMountain, quoted in
Freedom Voices Catalogue, 1999–2000

The concept of wilderness—long a central feature in American cultural discourse and a crucial element shaping our national ecologies—has not fared well in recent years. Noting the ways wilderness preservation functions as a primary concern in the mainstream environmental movement often to the detriment of other issues, such as industrial pollution, toxic dumping, labor rights, and occupational health and safety issues, grassroots activists involved in environmental justice have focused on redefining the scope of the movement, arguing that the world's most important ecological problems are invariably linked to other social inequities. In 1987 the United Church of Christ's Commission for Racial Justice published its report on the correlation between the racial makeup of a community and the incidence of toxic sitings. As the commission and other subsequent studies have repeatedly indicated, race operates as one of the most important factors determining decisions about where to locate hazardous waste.[1] Across the United States, communities of color disproportionately house the nation's toxic dumps; and in the Third World, the United States and other powerful nations continue to ship their industrial waste for storage even as they eye the resources of these less powerful countries. Given such realities, it is perhaps no surprise to note that for many environmental justice activists, wilderness preservation is simply not a primary political concern.

Environmental historians also have begun refiguring the debate, arguing for a recognition of wilderness as a fictive geography, a concept that primarily encodes Euro-American notions of nature, culture, and national identity. William Cronon, for instance, has gone so far as to suggest that the modern wilderness idea, rather than serving as a solution to our ecological woes, may instead be considered part of the problem. In positing a site of redemption away from our everyday living spaces, wilderness advocates frequently end up reifying divisions between human and nonhuman nature, setting up boundaries and borders that do not serve either entities particularly well.[2] Wilderness enthusiasts have often countered these claims, charging that such arguments play into the hands of developers by negating one of the environmental movement's strongest rallying cries.[3] Understanding the stakes involved in dismantling our dominant spatial mythologies, Cronon describes the fetishization of a particular kind of nature as a form of wilderness religion, a type of worship that has been central in the development of a Euro-American sense of self.[4] As wilderness religion shapes and defines the dominant national imaginary, Alaska continues to be positioned as a kind of Vatican City, a sacred site surrounded by a geography of profanity. The myriad ways in which Alaska has been mythologized in the U.S. spatial imagination as a wilderness site par excellence have in turn displaced and marginalized the concerns of indigenous people in the region. Environmental sovereignty for many Alaska Natives thus involves countering visions and uses of the land as a Last Frontier, an imperiled wilderness in need of protection or a pristine, unsullied terrain that time has somehow forgotten.

The attempt among Aleuts, Indians, and Eskimos to create different definitions of Alaska has resulted in another kind of political struggle, this time against certain forms of environmentalism.[5] Such battles intimately link Alaska Natives to the struggles of other minority groups in the United States and elsewhere, who also have found themselves at odds with mainstream environmental organizations. Winona LaDuke, Anishinaabeg activist and former vice presidential candidate for the Green Party, discusses the historical conflicts that often have prevented the development of long-term coalitions between native and nonnative environmental groups. She writes,

> [L]ines are often drawn between those environmentalists who can support Indigenous rights to self-determination and those who fundamentally cannot. Some call it environmental colonialism, others call it plain racism and privilege. The underlying problem is often quite basic, revolving around historic views of who should control land, perceptions of Native people, and ideas about how now-endangered ecosystems should be managed. Most disturbing is the widespread absence of any historic knowledge of traditional Native tenure on these lands and the demise of Native ecological and economic systems.[6]

LaDuke's comments remind us of the issues that typically don't figure in the larger environmental picture. Her observations raise the question about whose interests are at stake in determining the scope of the movement, observations that may be applied on a broader scale as well. As the globalization of capital radically transforms and reconfigures geopolitical boundaries, global environmentalism frequently operates as a new imperialism, with the forms of ecological interven-

tion that the West engages in often creating profound and quite negative consequences for the rest of the world.[7] Geographers Richard A.
Schroeder and Roderick P. Neumann speak of these new forms of
environmental intervention, describing what they call manifest ecological destiny. "What we are witnessing is the emergence of a new
sense of manifest destiny, a naturalized ecological mandate that drives
environmental organizations and their donors to assert control over
remote territories in new ways. As with the original manifest destiny
doctrine, environmental interventions are imbued with moral cer-
titude; supported by science, they purportedly restore balance and
equilibrium to a troubled planet."[8]

Becoming the world's new green police, policymakers in the United
States, equipped with the latest scientific knowledge, often assume that
they have better answers to the environmental crisis, disregarding local
knowledges and understandings of the natural world. Yet if the environmental crisis, as Joni Seager argues, "is a crisis of the dominant
ideology," then U.S. policymakers and environmental activists might
best be served by abandoning some of the old ways of thinking and
focusing on efforts to locate new ones.[9] This problem of ideology
appears dramatically as Euro-American definitions of nonhuman nature colonize other definitions of nature. In the case of wilderness
preservation, for instance, notions of nature as a stable, unchanging
refuge—or, as Jennifer Price puts it, a place "Out There," a "not-
modern Place Apart"—operate in direct contrast to Native American
understandings of the natural world.[10] Rather than serving as a place of
permanence located outside human domain, nature for many American Indians is instead understood as "intelligent, inventive, changing,
learning, teaching, evolving, acting, praying, feeling, and responding."
According to Jack Forbes, it is often considered dynamic and unpre-
dictable, and humans are not typically thought to be separated or
distanced from it. For many Native Americans, he explains, nature is
not "a passive, acted upon, place where only 'immutable laws' of
science operate. Instead, its essence seems to be an ability to modify
itself in response to new situations."[11] This nature, endowed with an
agency of its own and intimately linked to human culture, seems far
removed from the nature that wilderness advocates invoke, whose
value lies primarily in its status as a steadfast entity, its ability to offer a
sense of constancy and stability in a rapidly changing modern world.

If it is clear that we are dealing with different natures, with vastly divergent ideas and understandings of what constitutes the natural world, then it follows that we must no longer regard the environmental movement as a coherent entity, but must instead address what appear to be discrepant environmentalisms. This chapter is concerned with examining an overlooked element in the mainstream environmental movement, namely indigenous land rights. In the argument that follows, I trace forms of environmental thinking that have redefined popular understandings of Alaska and have countered forces of globalization and Americanization across the state. This thinking operates in stark contrast to the wilderness ideas of John Muir, Robert Marshall, Lois Crisler, and Margaret Murie, or the frontier ideologies of Jack London, Rex Beach, and James Oliver Curwood. Focusing on the work of three Alaska Native writers—Tlingit authors Nora Marks Dauenhauer and Robert Davis, and Athabascan writer Mary Tall-Mountain—this chapter investigates the ways indigenous Alaskans have envisioned an alternative environmentalism that dismantles popular constructions of Alaska as a Last Frontier, a wilderness playground for Euro-American adventurers, and a prime site for the United States' economic and territorial expansion. In doing so, I trace the ways indigenous populations across the region have also produced counternarratives that reenvision human relations with nonhuman nature and that require a sense of identity and reciprocity with the natural world, rather than a belief in human separateness from a wilderness Other, a vast Nature somehow located "Over There."

Recollecting Tribal Voices

Nora Marks Dauenhauer has distinguished herself over the past twenty years as a central figure in the movement to restore and reclaim Tlingit history, language, and land. An accomplished linguist, translator, biographer, editor, and poet, Dauenhauer, along with her husband, Richard, has published three volumes of work that address Tlingit oral history and culture. The first collection, *Haa Shuká, Our Ancestors: Tlingit Oral Narratives*, serves as an introduction to Tlingit social structures and oral literatures. The second volume, *Haa Tuwunáagu Yís, For Healing Our Spirit: Tlingit Oratory*, addresses Tlingit spirituality and worldviews, showing connections between verbal and visual art, ritual, and myth. The third volume, *Haa Kusteeyí, Our Culture: Tlingit Life Stories*, examines Tlingit social and political history, bringing together historical documents, personal photographs, and individual as well as collective memories. Dauenhauer has also been widely anthologized in various literary collections and has published two books of poetry.

Her first collection of poetry, *The Droning Shaman*, incorporates Aleut, Tlingit, Japanese, and European literary forms to capture the everyday details of her life as a writer, scholar, and activist. The collection *Life Woven with Song* continues this project and includes poems, prose pieces, and plays she adapted from the Tlingit oral tradition. Born in Juneau in 1927, Dauenhauer lived during her early years in seasonal fishing camps in Southeast Alaska. Her first language was Tlingit, with English entering her life when she began school at age eight. Like the two other native authors discussed in this chapter, Dauenhauer writes of the battles facing tribal peoples across the state as they struggle against cultural destruction and move toward cultural recovery. A self-described "born-again Tlingit," she has dedicated herself to reversing the history of cultural suppression across native Alaska.[12]

Dauenhauer's poem "Chilkoot River/Lukaax̱.ádi Village," for instance, describes the ways Euro-American development across Alaska has literally worked over and destroyed native cultures.

> After the bulldozer
> went through the village
> site,
> no one bothered to look
> to see
> if they had uncovered anything of
> significance.
> They left
> where they uncovered a grave
> site.
> The skull facing the road seemed
> to say
> "Look at me, Grandchild,
> Look at what is happening to me."[13]

Here landscape embodies Tlingit history, making it easy to see that for indigenous environmentalism, as Winona LaDuke argues, the main lesson is "that the war on nature is a war on the psyche, a war on the soul."[14] In this poem, the bulldozed landscape uncovers a buried Indian heritage that demands to be respected, preserved, and protected by the next generation. In this way, the past is not "history" or something set apart; instead, it is alive and connected to the present.

As a major theme in her life work and in her writing, cultural recovery underpins Dauenhauer's poetry collection as a whole. For her, the struggle is all about coping with what is available and transforming what is at hand into something useful. Thus, the poem "Mickey Mouse Comes to Juneau" tells about the invasion of the Disney empire into the state's capital city. Although Disney icons have come to represent global capitalism and commodity culture, the dregs of the American leisure industry, for the Tlingits and other residents of Southeast Alaska, Disney might also serve as a repository for other meanings. Such is the case for the Juneauites who head downtown to greet Mickey. For those people who may or may not ever venture to a Disney theme park in the Lower 48, Mickey has come to meet them. Dauenhauer writes,

> South Franklin was lined with kids
> from six months to forty years.
> It was the day attention switched
> from adult entertainment to children's
> on South Franklin's "Bucket of Blood." (DS, 17)

In the heart of Juneau, South Franklin was previously a run-down thoroughfare that recently underwent gentrification partly as a concession to the tourists arriving off the luxury cruise ships, whose main route to the shops and restaurants in downtown Juneau involved passing through the city's underclasses. Gentrification added some restaurants and gift shops, new landscaping and paint on the buildings, but didn't entirely succeed in changing the street's atmosphere.

This seemingly whimsical poem about the day Mickey came to Juneau, then, operates as a moment for celebration. Dauenhauer thus continues,

> Some waited three hours
> straining necks and eyes
> looking for the famous mouse
> . . . It was like
> lining up for sacraments. (DS, 17–18)

Mickey's arrival becomes an opportunity to imagine South Franklin for what it could be—not a site of social division but a place for all residents to gather as a community. And although it is the state's capital city, Juneau is renowned as a geographically isolated community. A constant source of irritation for the state legislature, for instance, has been the fact that the only means of transportation into the city are by ship or plane. With few outside forms of entertainment making their way into town, Mickey in this sense becomes a means of pleasure and a moment of possibility for Dauenhauer and other Juneau residents.

In "How to Make Good Baked Salmon from the River," Dauenhauer moves from the globalized icon of Disney to the local world of Tlingit history and culture. Using the form of a recipe, the poem provides its audience with a sense of how to prepare a basic Tlingit meal. Yet by including various cultural rituals behind the sharing of food, the poem also becomes something more. This seemingly straightforward recipe turns out to be a guideline for how Tlingits have understood themselves in the world and among others. She begins by explaining that the salmon is best made in camp near the beach and a stream using a wooden stick over the fire. Understanding that some of her Tlingit readers may not have access to these supplies, either because they are now part of a more urban environment or because they live outside Alaska, Dauenhauer writes, "In this case, we'll make it in the city, / baked in an electric oven on a black fry pan" (11). Taking into consideration the various conditions that might restrain her audience from following the recipe, she takes care to update the instructions for her current readers, thus expressing her fundamental belief that cultures are not lost but changed:[15]

> think about how nice the berries
> would have been after the salmon,
> but open a can
> of fruit cocktail instead. (*DS*, 13)

Dauenhauer later recommends finding some skunk cabbage by the stream to serve with the salmon ("because it's biodegradable"), but, given that many Tlingits no longer live by the river, "[i]n this case, plastic forks, paper plates and cups will do" (*DS*, 13).

Respect for the salmon and reciprocity with the natural world also operate as important elements in cooking salmon from the river. Thus Dauenhauer includes directions for ensuring that these concerns are part of the meal.

Gut, but make sure you toss all to the seagulls
and the ravens, because they're your kin
and make sure you speak to them
while you're feeding them. (*DS*, 12)

And because this meal initially was prepared in the wet rain forests of Southeast Alaska, she reminds her readers to "[s]hoo mosquitoes off the salmon" and to shoo the ravens away as well, "but don't insult them," she says,

because mosquitoes
are known to be the ashes of the cannibal giant,
and Raven is known to take off
with just about anything. (*DS*, 13–14)

Through this poem/recipe, the simple act of making a meal becomes the occasion for strengthening Tlingit cultural heritage. Later in her description of how to eat the salmon, she writes, "Everyone knows that you can eat / just about every part of the salmon, / so I don't have to tell you / that you start from the head." While this information might be elemental knowledge for her readers, she includes it anyway, since reminders never hurt. Being the devoted linguist that she is, Dauenhauer also slips in a language lesson, explaining, "You start on the mandible / with a glottalized alveolar fricative action / as expressed in the Tlingit verb als'óos." Finally, because the act of providing nourishment serves as an occasion for strengthening cultural ties, Dauenhauer writes that the meal should end with a storytelling session near the fire, but, in this instance, sharing small talk, jokes, and a beer with friends will suffice. "If you shouldn't drink beer," she suggests that "tea or coffee will do nicely" (14, 16).

In another poem, Dauenhauer presents Tlingit local knowledge, particularly as it clashes with metropolitan or "official" knowledges. In "Grandmother Eliza," she writes of the family surgeon who saved many lives with a scalpel made out of a pocketknife. Her grandmother's intern, Auntie Annie, was also a healer for the community. Dauenhauer remembers an accident she had as a child and recalls Auntie Annie searching in the night for the bark of plants she knew. Both healers are later warned by the white doctors, " 'You know you could go to jail for this?' "[16] Later her grandmother's other intern, Auntie Jennie, saves an uncle's life after his son accidentally shot him through the leg.

A doctor warned her, too,
when he saw how she cured.
Her relative cured herself of diabetes.
Now, the doctors keep on asking,
"How did you cure yourself?" (*LW*, 62)

Dauenhauer points out here with a certain degree of irony that the knowledge which once threatened white medical practices and was once cast as a form of primitive or barbaric healing is now sought out with great demand by the same physicians.

In her poetry, she also examines how the imposition of other knowledges and other cultural forces have threatened the integrity of Tlingit culture, focusing in particular on how global tourism has changed indigenous ways of life across the state. In recent years, for instance, travelers have been able to take chartered flights to tour Eskimo fishing camps in the Arctic, spending as little as half a day and a few hundred dollars for the opportunity. In the poem, "Village Tour, Nome Airport," Dauenhauer writes of a white tour guide with a "glossy smile" who "slides" over to a native seamstress and asks if she can touch her coat. She does so without waiting for permission, explaining to the tour group that they "make them by hand." The poem expresses a certain degree of self-consciousness about the tourist gaze, as Dauenhauer herself is positioned as an outsider and a participant in a travel industry that is disrupting the native community. In that sense, the poem questions dichotomies of colonizer and colonized. Later, Dauenhauer watches as the seamstress, "[a]s if frozen," sits in silence (*DS*, 27). Here the exotic Other—in this case the Alaskan Native woman—is recontextualized, reconfigured, and revalued in the white tourist economy as a new object of interest and wonder, an object to be touched and even sampled.[17] Although Dauenhauer does not address what the native seamstress might be thinking, the woman's silence nevertheless says much about the indignities of having her culture and her privacy invaded by these visitors. At the end of the poem, she writes of a youth in handcuffs waiting to be transported to California who also becomes incorporated into the tour as the guide knowingly tells her group, "They all go to jail" (27).

Dauenhauer writes, too, of the labor that shapes Tlingit relations with nonhuman nature, describing forms of work which establish a connection between the two that cannot be sentimentalized or romanticized. In "Salmon Egg-Puller—$2.15 an Hour," for instance, she writes of the exhausting work in the canneries:

> You learn to dance with machines,
> keep time with the header.
>
> . . .
>
> Grab lightly
> top of egg sack
> with fingers,
> pull gently, but quick.
> Reach in immediately with right hand
> for the lower egg sack.
> Pull this gently.
> Slide them into a chute to catch the eggs
> Reach into the next salmon.
> Do this four hours in the morning
> with a fifteen minute coffee break.
>
> Go home for lunch.
> Attend to kids, and feed them.
> Work four hours in the afternoon
> with a fifteen minute coffee break.
> Go home for dinner.
> Attend to kids, and feed them.
>
> Go back for two more hours,
> four more hours.
> Reach,
> pull gently.
>
> Go home for the day.
> Attend to kids who missed you.
> When fingers start swelling,
> soak them in Epsom salts.
> If you don't have time,
> stand under a shower
> with your hands up under the spray.
> Get to bed early if you can.
> Next morning, if your fingers are sore,
> start dancing immediately.

The pain will go away
after icy fish with eggs. (*LW*, 63–64)

Dauenhauer's poem offers a description of life for those Tlingits pushed into the cash economy of the fishing industry, detailing in dramatic ways the toll this work takes on the body and the family. In doing so, the piece becomes a powerful countervision to the spring advertisements that are posted on college bulletin boards across the Lower 48, which promise students a summer of adventure and high wages in Alaska. For Dauenhauer and the other Tlingit workers in the fishing industry, the job isn't part of a short-term wilderness excursion or a summer's visit to the Last Frontier. Instead, the labor that a salmon egg–puller does represents a job that pays the bills and puts food on the table.

In her poetry, Dauenhauer also addresses the dominant environmentalism that shapes Alaska, the imported foreign vision that encodes Euro-American understandings of nature as a place intended to be separate from human contact and connections. Her often-anthologized poem "Genocide," for instance, tells of the conflicts between subsistence living and wilderness protection. Using the form of Japanese haiku, Dauenhauer writes,

Picketing the Eskimo
Whaling Commission,
an over-fed English girl
stands with a sign,
"Let the Whales Live." (*DS*, 26)

Here the environmentalism of seemingly well-intentioned groups has unintended consequences for the indigenous peoples of the region. Aiding other projects of cultural suppression, it works hand in hand with the larger anti-Indian movement, a group with a diverse constituency ranging from white supremacists to Wise-Use activists.[18] Through the poem, Dauenhauer joins other environmental justice activists who point to the irony of placing blame for the globe's ecological woes on less powerful groups in the United States and elsewhere. As Vandana Shiva has argued, the green movement, which once grew out of local efforts to resist environmental problems created by globally powerful institutions, has long been co-opted by these same forces. She contends that as a result, "the 'local' has disappeared from environmental concern. Suddenly, it seems, only 'global' environmental problems exist, and it is taken for granted that only their solution can be 'global.' "[19] As environmentalism globalizes, multinational corporations free themselves from any responsibility, shifting the blame onto communities that have no global reach whatsoever.[20] The result, as Cynthia Hamilton explains, is that while the Third World and people of color in the United States generally have had little access to the economic benefits of development, "they are now asked to be partners in solutions to an environmental crisis created by others."[21] In a strange way, these groups are thus petitioned to give up material possessions that the privileged themselves still enjoy.

The narrowing of agendas and the creation of strange scapegoats has long been a problem within U.S. environmental organizations, especially with groups aligned with the philosophy of Deep Ecology. As Peter van Wyck points out, such groups often subscribe to a singular conception of the human, a movement that closes off "the possibility of heterogenous subjectivities by representing humans as a single ecological category."[22] This indifference to difference has the effect of making multinational corporations and American Indians, members of wealthy countries and those from less powerful groups, somehow equally responsible for the globe's environmental problems. A new "green imperialism" thus finds a moral cover for itself, as globalization is prescribed as the proper cure to environmental degradation.[23]

Ultimately Dauenhauer's meditations in "Genocide" reveal the new forms that imperialism takes as whale campaigns and other preservationist movements become forces contributing to Alaska Native dispossession and displacement. For these white wilderness advocates, Eskimo whaling is somehow aligned with larger industrial forces that threaten endangered life across the globe. Facing the loss of wild spaces and species, these preservationists make Eskimos into agents and symbols of ecological destruction. Rey Chow's discussion of "orientalist melancholia" may be particularly apt here. Chow draws on Freud's analysis of the melancholic as a person unable to come to terms with the loss of a precious object and who ultimately introjects this loss into his or her ego, directing the feelings inward. This figure differs from other kinds of mourners by engaging in a "delusional belittling" of the self and by feeling somehow unrightfully abandoned in the wake of loss. In the context of postcolonial theory, however, Chow rethinks Freud's analysis of the melancholic disorder, describing another response that isn't directed inward, but toward a third party who provides the mourner with a means of externalizing the loss and a place in which to redirect blame. Orientalist melancholia thus becomes a new form of mourning, Chow explains. What Freud once saw as a denigration directed toward the self now finds clear relief in the denigration of others.[24] In the case of much wilderness protection—of the form discussed in Dauenhauer's poem—the loss of nature experienced by Euro-Americans often becomes directed toward the racial Other, who in turn is made responsible for that loss, becoming a target of environmentalists' denigration and blame.

The problems that complicate mainstream wilderness and environmental preservationist organizations may be regarded as symptoms of a culture that forges strong divisions between nature and the social, as signs of a worldview that sets off certain spaces as sacred and pristine while leaving other places as sacrifice zones.[25] Such a vision contrasts with the understandings of nature/culture relations that operate in Dauenhauer's writing. Her piece "Seal Pups," for instance, again written in the form of a haiku, captures these interconnections. Dauenhauer describes the animal "[a]s if inside / a blue-green bottle / rolling with the breakers" (DS, 5). The division between human and non-human nature that shapes so much Euro-American thinking about the environment is not sustained in the poem. Instead, we see the boundaries blurred as the two seemingly separate worlds are enveloped by each other.

Raven and the American Dream

Published in 1986, Robert Davis's poetry collection, Soul Catcher, also addresses themes of cultural upheaval and cultural renewal. Davis, a Tlingit writer and carver who grew up in Kake, a small village in Southeast Alaska, writes of separations between family members, the political and cultural upheavals that shape Tlingit generations, and the fishcamps, the salmon, the spruce trees, the seaweed, and other elements that place him in a changing Tlingit world.[26] Like Dauenhauer, Davis tells of the environmental impact of white contact, of cultural displacement and cultural recovery. In his poetry, he writes of family members who were taken away from their homes and made to accommodate themselves to another culture. His collection chastises the American cult of progress and its effects on those considered "primitive." In turn, Davis's writings criticize the drive for museumification and the practice of collecting that remake Tlingit culture into something seemingly past and extinct.

In his poem "At the Door of the Native Studies Director," for instance, a Tlingit man speaks of his father, who was taken to a white school where he had the "old language" educated out of him by outsiders. Later, after the man is finally remade in their image and can meet their specific qualifications, he becomes the "native scholar" and is given a job in his own "corner office." In the poem, the son, addressing his father, remarks on the irony of how cultural authority becomes established in the white world:

Now you're instructed to remember
old language, bring back faded legend
anything that's left.[27]

Here, the colonizer's nostalgia for an authentic Indian belatedly tries to call back a lost world that it had earlier banished. As Davis indicates in the poem, however, this return to the authentic "native" is a return to an origin already invented by the colonizer and is thus not a viable return in the final analysis.

A similar critique appears in another poem that describes the impact of the tourism industry on Tlingit communities in Southeast Alaska, one of the mainstays of the region's economy. In "The Albino Tlingit Carving Factory," Davis writes of the ideology of consumption that radically transforms Tlingit artists' relationship to their artwork but that also ironically provides monetary support for them. The tourists who yearn to own a piece of the primitive past they believe they are encountering during their visits to Alaska seek carvings whose value they do not fully understand. Davis thus describes the necessary shortcuts the Tlingit artists take in their production of tourist arts. They do not spend time searching out the "perfect grained red cedar." They do not "talk to the trees" or properly season their boards. Instead, the materials they use come from "Spenard Building Supply" in Anchorage at "$2.15 per board foot" (23). These shortcuts serve as a form of resistance, a way of countering the encroaching commodification of Tlingit culture in the tourism industry. A century earlier, John Muir expressed his dismay at what he witnessed during his cruises through the Inside Passage, when visitors turned away from the majestic glaciers, directing their attention to the sides of the tour ships where Tlingits sold their wares. Muir's consternation arose because he considered the natural world of the glaciers, the mountains, the forests, and the sea more authentic than the cultural world of the Indians as presented in the tourist trade.[28] Like many Euro-American travelers in this period, Muir regarded Alaska Natives as largely corrupted by contact with whites and thus believed that the Tlingits he encountered on his cruises functioned as little more than a debased version of the "real thing."

This dichotomy between genuine and inauthentic Indians is a powerful white invention that the Tlingit carvers throw back on the tourists. If the white travelers typically feel anxiety about their encounters with the Native Other and have concerns that they may in fact be duped, that what they are buying may not be the genuine goods they desire, then Davis and the other carvers he describes exploit this opportunity. They give the tourists what they expect—the inauthentic tourist fare. Davis explains:

> We do not go to the iron-rich cliffs
> for red ochre paint mixed in stone
> paintbowl with dog salmon eggs
> spit through mouthful of spruce bark.
> Nor do we try for the subdued blue-green
> of copper sulphate with virgin's urine,
> or black from the deepest charcoal.
>
> You want crude carvings?
> You want them harsh and vicious?
> Okay,
> they might be African for all you care.
> Hell, we have to make a living too. (23)

Although ever anxious to encounter authentic native culture, the white tourists in Alaska are often ill equipped to distinguish clearly between the myriad cultures that they have deemed "primitive." For the visitor, in fact, a basic interchangeability exists among all so-called unmodern people. If it is a truism that the oppressed are required to know their dominators more intimately than their dominators know them, then Davis and the other Tlingit carvers he writes about may be said to have more than a strong acquaintance with Euro-American desires and yearnings. Delivering on what the buyers seem to expect anyway, the Indians at the carving factory give the tourists what they want. After all, they have to earn a wage for themselves in the new cash economy.

This critique plays out in another poem, as Davis foregrounds the ways Tlingits have struggled to survive the demands of Euro-American culture. In "Outgrowing Ourselves," he tells of the pressures that Tlin-gits face to become modernized, to rid themselves of tribal knowledge and adapt to white culture. He begins by describing the Great American Dream and the place (or nonplacement) of Alaska Natives in it. Davis writes,

Tell me how I need to get a respectable job
so we could get off the foodstamps and canned
fish and jarred berries, so we could get electricity
so we could get a radio, so we could get a C.B.
so we could get a TV, so we could get a stereo, so
we could get a vacuum cleaner, so we could get a
telephone, so we could get a typewriter, so we could
get a microwave oven, so we could get a truck
so we could get the hell out of here (34)

Cataloging the ideals of white America, Davis updates the nineteenth-century vision of that great cataloger Walt Whitman to include all the objects and items needed to become proper consuming citizens, the acquisitions and possessions required of the American Dream in the late twentieth century. For the Tlingit, the national dream means leaving the community, the village, the culture, and history. For the Tlingit, it means deciding "to go to college in a year or two / so we could learn we don't know proper things." These are all the require-ments for preparing "for the good life," he writes. For the Tlingit pressured to accept this life, however, sarcasm and irony may be a more useful response, as the "good life" promised here by consump-tion and mobility hardly seems like a dream worth having (34).

Yet Davis also remains aware that a retreat to some prelapsarian, precontact time is not necessarily an option in itself. In "What the Crying Woman Saw," he contends that

If all the clocks in the house stopped
the mirrors still would haunt.
We could not go back
to the past we read in your eyes (26)

The speaker explains the impossibility of returning to an older mo-ment of fishcamps and boardwalks in small Tlingit villages, of smoke-houses, of coves filled with tents, a time of campfires and handtrollers, of iron pots, of sweet red meat, an era before "cheap whiskey," when "we dipped our food in seal oil" (26). Eschewing the desire for an older, seemingly more authentic environment, a desire that often fuels white wilderness enthusiasts in Alaska, Davis ends his poem by asking,

> Did you imagine the darkest-eyed, brown-skinned boy
> one day stuck at a typewriter
> remembering time two different ways? (26)

In this poem, the postcontact present and the radical alternative of life another way exist side by side for figures like those described by Davis, as different senses of time and space end up colliding in one body.

Ultimately the presence of Raven, a trickster figure among Pacific Northwest Indians, serves to dislodge the problems imposed by the American Dream and the cultures of consumption that drive U.S. economies and ecologies. In his longer poem, "Saginaw Bay: I Keep Going Back," Davis writes of Raven, "[c]ocksure smooth talker, good looker," and tells of how it all began. Once Raven throws the light,

> everything takes form—
> creatures flee to forest animals,
> hide in fur. Some choose the sea,
> turn to salmon, always escaping.
> Those remaining in the light
> stand as men, dumb and full of fears. (14)

For Raven, the world is so remarkable that even he is amazed at its creation. In this piece, it is interesting to note that human beings do not operate as the primary ecological players. Instead, Raven, along with the forest animals and the salmon, remains a central aspect in the world, while men and women are thought to lack a similar knowledge and thus remain full of fear. This story of beginnings differs from the accounts associated with Euro-American origins and history in Alaska. In "Modern Indians," Davis, for instance, tells of a site near Kake named Hamilton Bay:

> This place some European captain named for himself so
> it could exist properly. (Go and subdue the earth and
> name everything in it). (22)

Davis remains clearly critical of the arrogance of American Adams who, in the spirit of possession and dispossession, take up the task of renaming the land, thus believing they have somehow called the world into being themselves.

In "Saginaw Bay: I Keep Going Back," Davis recounts the history of the region in the wake of cultural contact, building on traditional oral narratives to describe the arrival of whites.

> I've heard of men in black robes who came
> instructing the heathen natives:
> outlaw demon shamanism,
> do away with potlach,
> pagan ceremony,
> totem idolatry.
> Get rid of your old ways.
> The people listened.
>
> They dynamited the few Kake totems—
> mortuary poles fell with bones,
> clan identifiers lost in powder,
> storytellers blown to pieces . . .
>
> People began to move differently, tense.
> They began to talk differently, mixed.
> Acted ashamed of gunny sacks of k'ink';
> and mayonnaise jars of stink eggs,
> and no one mashed blueberries
> with salmon eggs anymore. (16–17)

Davis speaks of the changes thrust upon Tlingits from the outside and the effects these forces have on the ways Indians understand themselves and each other, the ways they comprehend their history and their relationship to the natural world. Ultimately, his response to cultural upheaval involves resisting the genocidal forces that threaten to annihilate Tlingit culture. One form that this reversal takes involves writing. In his discussion of Alaska Native poets, critic James Ruppert suggests that indigenous writers "face the somewhat paradoxical task of crying warnings and singing celebrations at the same time."[29] Ruppert notes in Davis a strong concern with bridging connections to the past and a commitment to adopt older forms of storytelling for contemporary contexts.[30] These observations help explain another stanza in the poem. Davis writes,

> The old ones tell a better story in Tlingit.
> But I forget so much
> and a notepad would be obtrusive
> and suspicious. I might write a book.

In it I would tell how we are pulled
in so many directions,
how our lives are fragmented
with so many gaps. (20)

Here the author uses poetry to bridge the past and the present, to make connections between himself and the community. He thus goes on to tell of one father, who as a young man was sent away to Sheldon Jackson, a Presbyterian industrial school in Sitka.[31] There he was supposed to learn to be a "[u]seful member of society," but instead returned a changed man. This handsome father was shy and sad but popular with the girls. "[L]ikeable," Davis writes, "[b]ut goddamn, you had to catch him sober / to know what I mean." Later, he explains that the effects of cultural dislocation travel down through the generations: "they say I remind them of him. / But you have to catch me sober" (18).

When a logging company from Outside moves into his community, the owners bring heavy cables which they erect from the beach to the woods, and later allow bulldozers to leave deep trails in the ground. The upheaval that the Tlingits have endured under Euro-American advancement takes both a social and an environmental form here. The Tlingit community is suddenly invaded by "[r]edneck rejects, tobacco spitters" who get drunk in their bunkhouses at day's end, harass the Tlingit women, and brag about the loads they carried, "who got maimed / and did they take it like a man" (18–19). For Davis, the loggers' violence is symptomatic of the larger disease of expansion and conquest: "Some men can't help it," he explains, "they take up too much space, / and always need more" (19). The logging company, which only provides a few jobs for the Tlingits, also leaves its marks on the land, violently gnawing away at the forest

till the sky once swimming with branches
becomes simply sky, till there is only
scarred stubble of clearcut
like a head without its scalp of hair. (19)

As native groups across Alaska continue to face difficult decisions about development and land use, many of the smaller communities such as Kake are hit especially hard, often facing a particularly invidious form of struggle that Robert Bullard has called "environmental blackmail."[32] These native communities, already facing a shortage of

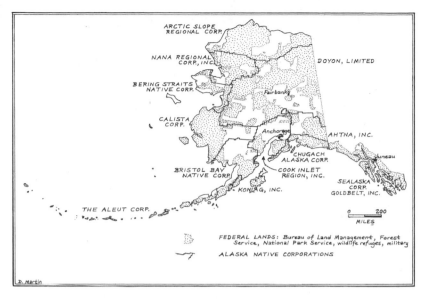

Federal and Native Corporation lands. (Map drawn by Dale Martin)

jobs, are often targeted by large corporations and multinationals as prime places for cheap labor and raw materials. For Tlingits who are struggling to enter the cash economy, the trade-off between traditional lifeways and economic opportunities complicates matters. In the wake of the congressional passage of the Alaska Native Claims Settlement Act (ANCSA) in 1971, whereby native land rights across Alaska were extinguished and replaced by the establishment of landowning corporations, the struggle for economic independence and cultural sovereignty has still not been settled. Some native corporations across the state have fared rather well, but others have faced bankruptcy. As several figures have pointed out, the corporate model introduced by ANCSA has brought with it new forms of "institutionalized competition" between native peoples that had not existed before and that violate a standard belief in forging reciprocal relations with the natural world.[33] Davis's collection of poetry aims to rewrite nature/culture relations from within a Tlingit belief system, dismantling along the way popular stereotypes that depict Alaska as a primordial wilderness region awaiting nature enthusiasts who seek to throw off a confining modernity in favor of a more authentic and primitive world. His work instead introduces the expansion of Euro-American culture and capital as central concerns for an indigenous environmentalism in Alaska.

Iñupiat writer Fred Bigjim describes the role of the native poet in cultural and political terms. "There are a multitude of problems facing Native people today," he explains. "There are attitudes and values of Native and non-Native cultures alike that poets can and should reshape. By making vivid what is at stake to both Native and non-Natives, our common American culture will be enriched, our sense of Nativeness will be enhanced, and the values of our society will be reshaped to accommodate positive change and action."[34] In many ways, Davis's poetry works in this manner, countering the political unconscious of mainstream nature writing that often emphasizes an act of retreat toward a world of solitude—a process of "losing the humans," as Randall Roorda so eloquently puts it.[35] Davis's poetry instead affirms the interconnectedness of all cultures and of human and nonhuman nature. In doing so, his writings become politically charged accounts of the histories of Tlingit resistance, not only to the cultures of expansion but also to the national ecologies that have altered and reshaped Alaska Native land tenure across the state.

Nomad and Native

The writings of Athabascan poet, novelist, and essayist Mary Tall-Mountain also foreground issues of environmental politics and Alaska Native sovereignty in particularly powerful ways. TallMountain was born in the small village of Nulato in 1918 to a Koyukon-Athabascan mother and an Irish American father. After her father left the family and her mother contracted tuberculosis, TallMountain was adopted by the Anglo doctor who treated the community, an arrangement organized by her birth mother which made TallMountain the first child in Nulato to be adopted out of the Alaskan village by whites. As she explains, the adoption was not greeted well by the rest of the community, and "[v]iolent revolt ensued" when she was taken away from Nulato.[36] At age six, TallMountain moved with her adoptive parents to the Lower 48. Because she had trouble coping with the vast changes created by the move, she and her family returned to Alaska, this time to the Aleutians, where her father found a post as a doctor. When TallMountain was fourteen, the family settled in central California. Through a series of bad investments, her father went bankrupt and died shortly thereafter when she was eighteen. A year later, Tall-Mountain's mother committed suicide, leaving her alone without any family support.[37]

TallMountain began writing seriously when she was in her mid-fifties, her work garnering notice in the late 1960s, primarily through the efforts of Cherokee/Chickasaw poet Geary Hobson.[38] She also was nurtured along the way by other writers, including Hopi/Miwok poet Wendy Rose and Laguna Pueblo writer Paula Gunn Allen, to whom she has dedicated many poems. TallMountain was working as a secretary in San Francisco when she met the latter poet, who was a professor at San Francisco State University. Allen later tutored her, and for a year and a half TallMountain wrote sixteen hours a day on a typewriter that soon became her "altar."[39] Over the years, she has published several volumes of poetry, and her work has been widely anthologized in various literary collections. After she began establishing herself as an important voice in Native American literature, Geary Hobson and other friends encouraged her to travel back to Alaska, a place she had not visited for more than fifty years. TallMountain finally did so in the mid-1970s, giving lectures, workshops, and a poetry reading at the University of Alaska at Fairbanks.[40] After her death in 1994, TallMountain's writings continue to be promoted and distributed by an organization associated with the Tenderloin Reflection and Education Center called the TallMountain Circle.[41]

Having lived in both urban and rural spaces, in the Tenderloin district of San Francisco for more than twenty years and in the Athabascan village of Nulato during her early childhood, TallMountain focuses on the experience of place and displacement and incorporates diverse languages into her writings, including Athabascan, English, and Spanish, in order to address her locatedness in the world. Many of her poems describe encounters with San Francisco's urban ecologies, while other poems recapture the spaces of her childhood in Alaska and tell of her experiences being removed from her people and culture. TallMountain's work is often haunted by memories of her childhood along the Yukon and is frequently dedicated to family members, particularly her birth mother, grandmother, and brother. Environmental health also operates as a central theme in her writings; over the years, she suffered three bouts of cancer and other ailments, which she largely attributed to the trauma of being separated from her people and her homeland.[42] Her work addresses other environmental concerns as well, linking the nation's postwar nuclear folly to Enlightenment ecologies of science and power, connecting conquest and genocide to themes of species extinction and the displacement of indigenous cultures and knowledges.

Most important are her own experiences of exile and dislocation which serve as a central concern in her writing, what Kenneth Lincoln describes as the working out of a "long homecoming."[43] In "Outflight," for instance, an autobiographical essay about her return to Alaska, TallMountain describes boarding a six-passenger plane, whose industry name "Nomad" aptly captures her own experiences being uprooted from her home more than fifty years ago (*LTW*, 6). In the piece, TallMountain ponders her identity as an outsider. Although she is an Alaska Native, in many ways she understands herself as an outsider traveling to a land that is largely unknown to her. In the essay, TallMountain ironically refers to herself as a "Cheechacko," a term used across the state to describe the newcomer (*TNW*, 7). The name, it is believed, originally derived from "Chicago," and referred among Alaska Natives to the settlers who sought their fortunes exploring, trading, and prospecting across the region during the nineteenth century. Residents in the state later picked up the word and used it to describe any greenhorn encountered in the North. TallMountain's essay tells of returning to her homeland, of reestablishing connections with her village and her culture, and of reversing her sense of dislocatedness in the world.

In the poem "Koyukons Heading Home," TallMountain speaks of exile again, beginning the piece by telling of her experiences sailing the Inside Passage along much of the same route John Muir took a hundred years before. But TallMountain's trip is different from Muir's. In this instance, she is not traveling as an eco-tourist in search of wilderness adventures; instead, in the poem she is interested in coming to terms with her own history and is heading to a place that was once her home. She encounters other Koyukons on the ship who are similarly displaced from their homeland, having had to leave their villages in search of educations or jobs. Watching them return, TallMountain takes some comfort, realizing that there are people who share her experiences and is encouraged that they, too, are heading home to their communities and their families (*LTW*, 48).[44] Ironically, on the day she was to travel to Alaska for the first time since she was a young child, TallMountain located her estranged birth father, now an eighty-four-year-old former musician and writer who had been living in Phoenix. Although she continued her trip to Alaska, afterward she flew to Phoenix to see her father and stayed there with him for the last two years of his life, nursing him as he struggled with cancer. Having experienced cultural displacement and feelings of abandonment much of her life, Tall-

Mountain later described the time she spent with her birth father as an opportunity to start the process of forgiveness and healing.[45]

In "Drumbeats Somewhere Passing," TallMountain positions herself once again as a nomad, a figure who yearns for the lost world and lifeways of her people. The poem addresses TallMountain's desires to connect to the culture from which she was taken as a child even as she is fully surrounded by elements of her adopted culture:

> serene I sip
> chamomile tea from English bone china
> yet by strange legerdemain
> of mind I feel
> drumbeats somewhere
> passing
>
> . . . just a breath away
> is another space
> where I ought to be
> I could almost flick a button
> and find it (*LTW*, 73)

Here TallMountain is encompassed by signs of a settled and serene domestic life yet nevertheless speaks of another world that comes to her through the sound of ritual drumbeats. This other life, appearing to her as a distant memory, is a place she "ought to be," a place of family and community. This other place contrasts with her lifelong experiences as an outsider, someone thrust violently into a foreign world and forced to survive or perish. In another poem, "Indian Blood," for instance, TallMountain tells of being the only Alaska Native growing up in her community in central California, of enduring other children's racist taunts. At her introduction to other students in her new grammar school, TallMountain was placed on stage where she recalls being dumbfounded by their reaction. Looking out at the white faces in the audience, she realized that she was perceived as an object of display to them.[46] "Do you live in an igloo? / Hah! You eat blubber!" TallMountain reflects on the anger she learned to direct inward:

> Late in the night
> I bit my hand until it was
> pierced
> with moons of dark
> Indian blood. (*LTW*, 26)

Throughout her writings, the process of reconnecting with her people and reliving memories of her family enable TallMountain to reconfigure the meanings attached to Indian blood. In fact, it was precisely because of her career as a writer that she was given an opportunity to finally return to Alaska and make connections with her village.[47]

TallMountain's poem "The Light on the Tent Wall" builds on the yearning to belong, and in doing so, describes displacement once again in stark terms. Dedicated to her birth mother, the poem imagines her own experiences as an unborn child waiting inside her mother, a figure whose round belly is reflected on the light of the tent wall. Later in the piece, TallMountain describes being wrenched violently from these connections to her birth mother and homeland:

> Years came. I was taken
> where there were no tent walls,
> where I had to dream my own
> and as time passed, often
> I saw the light on the wall.
> No longer pink, it was
> fire, its tongue licking
> the tent wall.
> Fire of our life, flickering. (*LTW*, 18)

By the end of the piece, TallMountain tells of rejoining her people and finding a means of talking about these experiences. The process of forging such connections, however, requires a coming into voice, which TallMountain soon discovers, is full of difficulties. She explains, "Often the sound was angry, / hasty, wanted to speak / but could not find words" (18). In a related way, Seneca social worker Agnes Williams speaks of what she calls "ethnostress," a term she coined to describe "what you feel when you wake up in the morning and you are still Indian, and you still have to deal with stuff about being Indian— poverty, racism, death, the government. . . . You can't just hit the tennis courts, have lunch, and forget about it."[48] The process of coping with ethnostress is a central element in TallMountain's work and has long operated as a motivating factor in her decision to keep writing (39). For TallMountain, the history of Euro-American encroachment in the Far North, which led to the spread of deadly contagious diseases, the resulting deaths of her mother and brother, and her experiences of displacement and exile, all function as instances of an overwhelming racial stress and burden that have directly fueled her writings over the years.

The stresses of racism have also led TallMountain to meditate on other topics, on individuals' and communities' struggle for environmental health and on the larger aims and actions of the mainstream U.S. environmental movement. Many of her poems thus focus on the concerns American Indians share with the animal world and address human relations with nonhuman nature. Like communities of color elsewhere in the United States who are often located outside the national ecologies that govern the dominant environmental imagination, the animal world faces a dire future which TallMountain addresses at length in her poetry. In "O Dark Sister," for instance, she writes of a discovered whale whose fate mirrors her own history of displacement and relocation. TallMountain begins the poem with a report that appeared in the *San Francisco Examiner* in the fall of 1979; it tells of curious motorists who stalled traffic as they struggled to watch a group of marine biology students recover a blue whale that had been beached near Pescadero. In the body of the poem, TallMountain speaks to this "dark sister," highlighting the affinity she and other native people share with the blue whale.

What instinct
Magnetized you to shore
Among plastic trash and rotted fruit,
Offal of careless creatures
Who so lately found the sands
Of your millennial home.
What dim kinship
Called you here? . . .

In a museum stand wired
The striding bones of Allosaurus,
Jaunty at his neck a plastic ruff.
Shall we someday see Great Blue
Daubed in dayglo green
When her bones are assembled
Like his, for the mere amaze
Of some unborn generation? (*LN*, 18–19)

Like other Indian nations whose millennial homes have been invaded and occupied by white America, this whale finds herself forced to relocate to new space. And like the poet herself, the whale must search for a new place to call home. Here TallMountain ponders the life of the animal whose endangered status is tied to a larger history of Euro-American encroachment that has resulted in Indian dispossession and genocide across the continent. Although mainstream environmentalists clearly understand species extinction as a central issue in their movement, TallMountain indicates that for native peoples, extinction not only marks the deaths of brothers and sisters in the animal world, but also serves as a reminder of their own extinctions and near extinctions from the time of European contact to the present era. In this way, TallMountain's poem recalls Melville's famous whale and the author's sweeping survey of Euro-American impact on nonhuman nature. Here, however, TallMountain adds another layer of analysis to these events, indicating the ways some human beings see themselves and their histories as intimately connected to other species endangered by the expansion of capitalism and nation building.

> Ask the bones that dangled in carnivals;
> Ask the bones traded by voyageurs;
> Ask the bones of Kintpuash and Black Hawk;
> Hear them at Elk Creek:
> Hear them at Sand Creek, Wounded Knee.
> Hear the ancestors' ghostly cries,
> Hear fabulous buffalo
> Begging back his giant bones
> Out of the carpeted plain. (*LN*, 19)

In a similar fashion, Winona LaDuke argues for a less disjointed understanding of human relations with the natural world, insisting, like TallMountain, that a direct relationship can be made across the Americas between the loss of biodiversity and the loss of cultural diversity.[49] Here the fate of the whale mirrors the fate of Native North America; hunted to extinction, the whale now symbolizes a world lost, a world that Euro-Americans mourn for and often attempt to recreate in zoos, museums, and other institutions dedicated to commemorating the dead. Having killed off the "primitive" to make way for "progress," white Americans later mourn this loss and present themselves as victims tragically bereft of a valuable object, adopting a pose of innocent yearning that obscures their own complicity with this history of domination. In the wake of destruction, they belatedly establish institutions to lament the very thing they earlier sacrificed.[50]

TallMountain develops these interconnections between the fate of Indian peoples, the threats to endangered species, and the drive for preservationism in her poem "Once the Striped Quagga." In the piece, she describes herself as a potential figure of extinction, given the history she has lived as an adopted Athabascan child of a white couple. TallMountain writes,

Look upon my face.
Its like shall soon be gone:
Flotsam of yet another race . . .

Once the striped quagga lived,
And the tender hyrax
Populous as Bengal Tiger,
Princely golden cat whose destiny
Hangs in the scales with ours:
Trees, beasts,
Other life-things who will
Inescapably surrender.

Sever the flesh from my bones.
Hang them above a fireplace,
Frame the mounted head
In arctic fur
Or exotic plumage
Such as is seen only in zoos or
Left captive in rapidly dwindling
Rainforests. (*LN*, 56–57)

Clearly, for TallMountain, endangered regions such as the declining rain forests and extinct species such as the striped quagga or the Bengal tiger share a similar fate with the indigenous peoples of the Americas: they, too, have become victims of the cultures of expansion and the subsequent cultures of collecting that shape and inform Euro-American history and identity. In the poem, the centuries-long mismanaged environments of white America are the end result of the nation's expansionist yearnings. In this sense, even the treatment of Alaska as a Last Frontier, a wilderness region at once precious and preserved, testifies to the ways the cultures of expansion lead to a deadening culture of collection for entities once considered in the way of progress but now revalued as precious and rare objects.

In her other poetry, TallMountain makes further connections across worlds and cultures, across histories and places, across the lines that divide Indian nations from one another. She dedicates a series of poems, for instance, to the crises shaping northern California's urban ecologies and describes the ways race has historically informed and complicated the nation's environmental thinking. Describing the devastated indigenous populations of northern California, TallMountain remembers the forgotten Indians who once lived in the area around present-day San Francisco. In the poem "Listen to the Night," she writes of a time "before land-fill," of a time when one could hear the "ancient / deer songs of hunters" and the "forgotten cries" of the bird populations now forced into extinction by the expansionist practices of white America.

> Listen, I hear
> many murmurs of a long-gone people,
> Ohlone Indian families
> lived in neat low tule tents,
> round, well made, in rows
> between fertile marshlands where now
> condominiums throw tall dark shadows (*LN*, 46)

Like the Athabascan world of her childhood that comes back to her in the form of a haunting memory, the forgotten history of the Ohlone Indians, long sacrificed to the onward movement of progress, refuses to stay buried in the past but emerges to disrupt the official memories of a city famous for its progressive politics, radical environmentalism, and racial tolerance. In the poem "Voices from Isla De Alcatraces," TallMountain likewise restores to memory an alternative environmental and racial history, telling of a place

> here where the pelican danced
> once known as *alcatraces*
> we see the seven hills
> cloaked in a dazzle of light
> boatloads of people
> staring at our sign
> YOU ARE ON INDIAN LAND
> laughter blows faint on the wind
> . . .
> we would cleanse the island
> of concrete and steel

let the healing grasses creep
let the creatures return
you bulldozed that plan
like everything else
America
for the sake of your furious fatness. (*LN*, 48)

Unmoved by the scorn directed toward them, undaunted by the sheer difficulty of their task, the poem's speaker and the other Indian protestors who are present insist on reclaiming a different history of San Francisco and Alcatraz, a different kind of land use and land tenure that challenges the wisdom of how Euro-American cultures have lived with nature and the ways they have fashioned urban spaces and ecologies.

Animal knowledges and the fate of America's urban landscapes also come together in a series of writings that center around the nation's nuclear future. These protest poems tell of a post-apocalyptic world where the only survivors are a lone wolf and an Indian woman who befriend each other in the wake of disaster. In the poem "The Last Wolf," TallMountain writes of the wolf, long a potent symbol among European cultures for the dangers of the wild and long respected by many Native American cultures for their intelligence and hunting skills. She traces the animal as he moves through the ruins of a city in the aftermath of nuclear war, searching for the only surviving human, an Indian woman who lies recovering in a hospital bed. In the piece, the woman hears his "baying echoes" as he moves through the devastated city center, its overbuilt landscapes now rendered useless, signs of a culture that placed tremendous faith in Enlightenment science but now at the edge of annihilation (*LN*, 62). Making his way through the demolished high rises with their elevators left useless, through the red and green traffic signals that blink indiscriminately, and through the rubble and debris of empty, quiet city blocks, the wolf

. . . trotted across the floor
laid his long gray muzzle
on the spare white spread
and his eyes burned yellow
his small dotted eyebrows quivered

Yes, I said.
I know what they have done. (*LN*, 62)

The wolf and the Indian woman connect with each other in this apocalyptic American setting, speaking each other's language and sharing a knowledge of the underside of the nation's ecologies of progress.

From her poems and essays about exile and displacement to her writings on the histories of U.S. expansion and progress, TallMountain's writings extend definitions of environmentalism and environmental health, reminding us of other knowledges that often remain hidden from the dominant culture, other ways of being in the world that have frequently been overlooked. The culture of expansion from east to west, north to south, and south to north, the ecologies of progress that sacrifice the "primitive" to make way for the new only to later lament the loss, the racialized stresses of exile and of resisting the dominant culture's environmental ideologies, are all aspects of a larger counterenvironmentalism that TallMountain and other American Indian writers and activists are striving to create. To make this claim is not to argue for an understanding of Indians as America's original ecologists, nor does it romanticize Native Americans as people who are collectively somehow closer to nature than European Americans.[51] LaDuke addresses this concern in ways that are useful to my discussion here, arguing that the "broader environmental movement often misses the depth of the Native environmental struggle. Although it has been romanticized historically and is often considered in some New Age context, the ongoing relationship between Indigenous culture and the land is central to most Native environmental struggles."[52] In this way, the argument TallMountain's poetry puts forth helps restore the question of power to our ecological debates and introduces issues of expansion and conquest into larger environmental conversations. In doing so, her work reminds us of the ways environmental issues are often constituted and how certain environmental concerns and forms of knowledge are made to matter over others.

To some extent, the use of TallMountain's writings in a discussion of environmental politics and Alaska Native sovereignty opens up a series of complicated issues that need to be addressed. As I've mentioned previously, the voice speaking in much of her poetry is one that has been shaped by specific social and historical conditions which dramatically changed the course of Alaska Native life in the twentieth century. Having been adopted out of her village as a young child, TallMountain expresses a strong desire to reconnect with her past and her community in many of her writings. There are some complex power dynamics that arise, however, in situating TallMountain as a spokesperson for Alaska or Alaska Natives, and in my discussion here, I have tried to avoid positioning her in the role of the urban Indian returning to the "traditional" community with unsolicited advice and all the answers. In that sense, it is important that TallMountain is not made to speak for all of Alaska or for all Alaska Natives. Her writings, however, operate as an expression of hybridity and open up space for rethinking notions of authenticity. In particular, they enable us to further dismantle ideas of a natural Alaska and to continue questioning essentialist understandings of indigenous peoples as pure primitives. The oppressive history that TallMountain grapples with in her writings precludes any attempts to situate Alaska as a site of pure nature or to regard the state's native inhabitants as an isolated people frozen in time, somehow removed from events that have shaped environments and populations elsewhere in the United States.

Indigenous Environmentalism and Alaska Native Sovereignty

As the poetry of Dauenhauer, Davis, and TallMountain indicates, indigenous environmentalism in Alaska has developed a different focus and form that distinguish it from mainstream U.S. environmentalism and the national ecologies that shape popular responses toward the region. From the myriad uses to which the region was put from the time of purchase to the present era, Alaska Natives have responded to U.S. territorial encroachment and globalization by developing an indigenous movement that addresses alternative land-use practices and that foregrounds issues of native sovereignty across the state. As the writers discussed in this chapter demonstrate, mainstream U.S. understandings of nature and ecology that underpin dominant conceptions of Alaska cannot be separated from deep-seated cultural myths, from ideas of expansion, and from larger nation-building projects. In many ways, then, mainstream environmentalism for many Alaska Natives is not necessarily the answer but frequently operates as a problem to be reckoned with in its own right.

Native environmental sovereignty in Alaska likewise involves efforts to counter popular visions of the land as a wilderness or a Last Frontier, an anachronistic space awaiting Euro-American arrival or an unclaimed terrain to be used for purposes of self-glory and self-promotion. Alaska Natives have often envisioned different ways of understanding the region, creating images and portraits of a land where human beings forge respectful and reciprocal relations with nonhuman nature, and where the divisions and distinctions between nature and culture are not so deeply defined. As environmental scholars and activists have increasingly argued, it is becoming clear that many of our current ways of thinking about wilderness are at a dead end and even pose danger to human and nonhuman nature. If environmental advocates want to forge changes in the ways nature should be conceived, then we must consider seriously an ecology that integrates human beings into nature and that does not set up certain lands as sacred spaces while condemning others as sacrifice zones.

TOWARD AN ENVIRONMENTAL CULTURAL STUDIES

The environment [is not] an ontologically stable, foundational entity we have a myth *about*. Rather, the environment is *itself* a myth, a "grand fable," a complex fiction, a widely shared, occasionally contested, and literally ubiquitous narrative. . . . "[T]he environment" and "wilderness" are effects of environmental discourse. . . . Such things become "natural" to us only through the reiterative work of the discourses that deal with or speak for them. The natural environment seems natural only because it is continually reconstituted as such, on a daily basis and at seemingly benign levels of perception, within interlocking systems of signs generated and perpetuated by living institutions.
—David Mazel, *American Literary Environmentalism*

Ideas of nature never exist outside a cultural context. . . . Nature as essence . . . as naïve reality . . . wants us to see nature as if it had no cultural context, as if it were everywhere and always the same. . . . If we wish to understand the values and motivations that shape our own actions toward the natural world, if we hope for an environmentalism capable of explaining why people use and abuse the earth as they do, then the nature we study must become less natural and more cultural.
—William Cronon, "Introduction," *Uncommon Ground*

If Alaska has been popularly imagined as a state of nature, the nation's Last Frontier, and an imperiled wilderness in need of protection, then the responses of Joel Fleischman from the critically acclaimed television series *Northern Exposure* come close to representing a truly alternative vision of the region. As one of television's most famous displaced young urbanites, a Jewish doctor from New York who has been exiled to the Far North as a means of paying back his medical school loans, Fleischman describes Alaska as "somewhere between the end of the line and the middle of nowhere."[1] Situated in the mythical town of Cicely, Alaska, in the 1990s the popular screwball comedy exploited the character's misery from one season to the next. In episode after episode, Fleischman makes no qualms about the cultural deficiencies of this Last Frontier, a place whose residents have little or no knowledge of the concept of bagels and whose relations with the world out-of-doors are a bit too intimate to his liking. In the show's pilot, which first aired during 1990, the character spends only a few hours in Cicely before he desperately declares his need to get out, protesting his fate in having to spend the next four years of his life in a "godforsaken hole-in-the-wall pigsty with a bunch of psychotic rednecks."

In each episode of the CBS series, Fleischman is consoled by other characters in the show, including the aspiring documentary filmmaker and mixed-blood Indian, Ed Chigliak, who relies on various cinematic references to make sense of the doctor's experiences in Alaska. During one episode, Ed comments on the ways Fleischman's suffering reminds him of the film *To Sir, with Love* (1967), where Sidney Poitier's character works in an inner-city London school with a group of underprepared, disaffected teenagers. Throughout the film, the character continually laments his "urban wilderness" surroundings until he eventually has a change of mind. "At first he hates it, but then he gets to know his students and care for them dearly," Ed explains in his typical deadpan delivery. At another point, Ed compares Fleischman's four-year term in Alaska to James Cagney's prison sentence in *White Heat* (1949), which features a finale with the psychopathic Irish gangster going out in a blaze of fire atop a gas storage tank in a machinery building several stories from ground level, hollering, "Top of the world, Ma! Top of the world!"

While Fleischman considers his own stay "at the top of the world" as a terrible form of torture, the rest of the characters in the show respond to the Alaskan setting in an entirely different manner, blissfully going about their lives and facing only minor conflicts or in-conveniences. Mainstream definitions of success do not carry much weight for these characters. In one episode of the show, for instance, the shopkeeper Ruth-Anne expresses disappointment after the visit of her son Matthew, a highly successful investment banker on Wall Street. "He isn't like you or me," she confesses with some embarrass-ment to her friends. As it turns out, Ruth-Anne prefers the company of her other son, Rudy, the truck-driver poet who doesn't waste his whole life "maximizing the returns" on his investments, but instead shows more interest in fulfilling his own human potential. *Northern Exposure* proved popular to U.S. audiences at the time not only be-cause it declared itself different from everything else on television and was broadcast during an era when the networks were touting what became known as "quality television." The show also proved popular to audiences because it focused on an area of the country that had recently received a great deal of media attention (it first aired only a year after the *Exxon Valdez* oil spill) and featured images of a breath-taking landscape that had been largely missing on network television shows (it was filmed, though, in Washington State, not Alaska).[2] At the same time, *Northern Exposure* departed from the usual format, offering itself up as a different interpretation of Alaska, doing so in a manner that featured rather unexpected scenarios and characters as well as some truly quirky views of nature/culture relations.

Throughout the series's run, for instance, the world out-of-doors continually invades the cultural realm occupied by the characters. In the opening credits of each show, audiences are given a glimpse of a moose sauntering across an intersection in downtown Cicely. In the series's premiere episode, Fleischman finds himself during his first day on the job in the strange position of caring for a sickly pet beaver who has been brought to him by one of his new patients. Later in the episode, he looks out his truck window and sees a group of people inexplicably communing with a llama on a downtown park bench. Perhaps one of the show's most clever accomplishments, however, appears in the way it often recast popular ideas about Alaska as a promising Edenic landscape. *Northern Exposure*'s critical take on the fetishization of prelapsarian nature emerged most notably in its treat-ment of the characters Adam and Eve, the constantly bickering two-some whose petty quarrels regularly spill over into Cicely's daily life. The show's Adam and Eve are featured as a woefully discontented, dysfunctional husband-and-wife team. Adam is played as a belligerent, verbally abusive misanthrope who stops just short of being physically threatening as well, while Eve is portrayed as a self-absorbed hypochondriac whose relations with the outside world always prove disastrous. At the same time, both characters are figured as highly capable and accomplished individuals: Adam was once a gourmet chef in New York City whose specialties ranged from Szechwan cooking to Northern Italian cuisine, while Eve possesses a deep familiarity with the mainstream medical establishment as well as a remarkable knowledge of rare, largely unheard-of illnesses and diseases. Fleischman's initial encounter with Adam, the grizzled Vietnam veteran who has just spent the last fifteen years of his life in Alaska's backcountry, offers new insights into the problems of wilderness and the American yearning to locate a prelapsarian garden in Alaska. The impulse to transform oneself into an American Adam or an American Eve carries huge costs, the show seems to indicate. Be careful what you ask for—a return to the garden may result in a community of caustic and belligerent Adams and Eves whose antisocial behavior promises to make for a very difficult Edenic existence.

While the show added a new twist to the "greening" of television in the 1990s, it also participated in a renewed cycle of the Western that was set in motion with the appearance of films such as *Dances with Wolves* (1990).[3] Just as *Northern Exposure* played with audience's expectations about the natural world, the show also reconfigured the popular Western, opting for a certain degree of self-reflection toward the genre and its codes. As an ironic send-up of the form, *Northern Exposure* features all of the conventional character types found in the genre. The neurotic, hyper-urban doctor, Joel Fleischman, for instance, serves as the eastern tenderfoot constantly confronted with unusual situations and even stranger locals. Maggie O'Connell, the privileged suburban daughter of Grosse Pointe, functions as the plucky and capable frontier woman, in this case a bush pilot who provides taxi service across the state, delivering mail to isolated communities in the Far North. Meanwhile, Maurice Minnifield, the former NASA astronaut, serves as the frontier booster. A landowner and powerful businessman, he plays the classic town promoter who believes that if you build it, they will come. As he confidently boasts to Fleischman in the first episode of the show, it "may not be today, may not be tomorrow, but it's coming. I guarantee you that." *Northern Exposure* also updates other classic Western types, including Holling Vincoeur as the town bartender, Shelly Tambo as the saloon girl, and Ed Chigliak and Marilyn Whirlwind as the show's representative Indians. Yet even as the series borrows from the codes of the genre, the characters only provisionally inhabit the traditional roles of the Western, and the Alaska that *Northern Exposure* depicts largely figures as an irreverent, anti-Western space.

One show-within-a-show episode featuring the founding of the town, for instance, reconfigures gendered conventions of the Western in rather unexpected ways. According to an old-timer who lived in Cicely at the turn of the century, the town started out as a squalid, lawless place until two young women, Cicely and Rosalyn (who are also lovers), appear with changes in tow. Revising the codes of the Western, which dictate that the arrival of white women signals the appearance of "civilization" in the region, *Northern Exposure* questions the truism that such women will necessarily bring positive social changes to the land. When we are introduced to this northern settlement, for instance, there are already numerous white women in place, including an intrepid Salvation Army missionary played by Maggie O'Connell. Although the arrival of white women falls short of ensuring that a certain form of culture will survive on the frontier, the show playfully updates the codes, suggesting that instead, the appearance of lesbians helps secure a kind of "civilization" in the region.

As independent and capable New Women, Rosalyn and Cicely work hard to clean up this frontier community. They bring the first automobile to the region, build an artists' utopia that showcases the work of various writers and musicians, and even lure Franz Kafka, who begins to recover from a bad case of writer's block during his stay in Alaska. The two women come close to taming the town outlaws as well, laying bare their acts of violence as nothing more than the misplaced behavior of children suffering from some unresolved childhood trauma. "It's acceptable to have feelings of anger," Rosalyn tells the two outlaws, "but it's unacceptable to act on those feelings. . . . Children lash out when they're angry. We're adults. We can talk." Cicely in turn adds her thoughts, telling them "we understand that you have this need to hurt people. . . . Someone must have treated you very badly when you were a child." The two women nearly succeed in disarming the men with their strange therapeutic talk, but in a gunfight that ensues, Cicely receives a bullet in her chest and dies in her lover's arms. The community eventually learns from the example of the two women and finishes the job of running the men out of town themselves. The show closes with Rosalyn, still mourning the death of her lover, leaving Alaska for Spain where she spends the last years of her life fighting Franco and his fascists.

In addition to reconfiguring gender roles in the Western, *Northern Exposure* also tries to recast racialized conventions of the genre. The character Ed, for instance, works against the popular image of the tragic and doomed Vanishing American. A rap-loving documentary filmmaker and shaman-in-training, he proves to be just as urbane and city-smart as Fleischman himself. Although the producers of the show, Joshua Brand and John Falsey, once commented that they envisioned Ed as a figure who had traveled no more than a fifty-mile radius of his hometown,[4] the character nevertheless possesses a remarkable amount of cultural capital. Clad in a hip leather biker jacket, Ed is an avid fan of *St. Elsewhere* (also the brainchild of creators Brand and Falsey) and has the ability to quote film history at the drop of a hat. When Fleischman shows his surprise that Ed is familiar with life in New York City, Ed explains, "I saw *Manhattan*. I think Woody's a genius." At another point, Ed befriends a German techno-punk art dealer named Rolf Hauser, who stands on a kitchen table one evening and recites dialogue from *Dances with Wolves*. "You Indians are so visceral," he announces earnestly. "Thanks," Ed responds, and then comments that he, too, has seen a lot of movies about Germans, including *Marathon Man* with Laurence Olivier as a Nazi doctor. Turning the tables on his German admirer, who receives most of his information about Indians from Hollywood film, Ed asks Rolf, "How does it feel to be always the bad guys? I mean . . . in the movies." In another episode, Ed suffers from director's block and has a dream in which he is presented a prestigous award at the American Film Institute. During the ceremony, he shakes hands with old friends Spielberg and Fellini, the latter eventually making his way to Cicely (the cold one, not the hot one, we are told) to visit Ed on his home turf.

Although it focused on contemporary social problems that usually went unspoken and unexamined in the Western, some of the show's critiques were not always fully successful. At times, for instance, *Northern Exposure* took shortcuts assembling bits and pieces of various American Indian traditions as if there were no perceivable differences between them worth noting.[5] A few years after the show ended its run on CBS, Cheyenne/Arapaho director Chris Eyre felt the need to comment on the series by recasting some of the characters from *Northern Exposure* in his film *Smoke Signals* (1998). Eyre, for instance, used actress Elaine Miles, who played Marilyn, in a wonderfully subversive part that proved to be a far cry from the side-kick role she played opposite Rob Morrow in the series. Cynthia Geary, who played Shelly in *Northern Exposure*, also was also cast in Eyre's film, this time as a gymnast whose unself-conscious, never-ending cheerfulness revealed a lack of depth and sincerity. Chris-in-the-Morning was likewise parodied as Randy Peone, the radio personality that replaced him in *Smoke Signals*, a figure who didn't try to promote the small town in a hyped-up morning monologue but told it like it was. Although it seems that there was nothing much happening for him to report on air, the DJ nevertheless opened his show at one point by announcing, "It's a great day to be indigenous."[6]

The producers of *Northern Exposure* also made a telling remark early in the show's run that they envisioned the relationship between Joe and Ed as a kind of updated version of *Robinson Crusoe*,[7] thus transposing a colonial narrative onto their seemingly revisionist treatment of Alaska. The theme song to the show played into this vision; with its quasi-Caribbean opening music, *Northern Exposure* contributed to a certain extent to the myth of the Last Frontier by positioning Alaska as a remote natural getaway awaiting the arrival of leisured Americans. At times, *Northern Exposure* also seemed driven by a touristic promotional ethos, albeit with a certain degree of irony. As one Alaskan bureaucrat tells Fleischman in the show's first episode, "Once you have experienced Alaska and I mean the *real* Alaska, everything else pales in comparison. Aspen's got nothing on this place." *Northern Exposure*'s appeal in this regard was not lost on the movie industry after the show's first airing. Ann Fienup-Riordan notes, for instance, that the popularity of *Northern Exposure* eventually gave rise to what she calls an "Alaskan location rush" in the 1990s, as moviemakers interested in locating an exotic new American setting discovered the possibilities of filming Alaska.[8]

Finally, although *Northern Exposure* worked hard to stage moments in which Indians revised the roles typically assigned to them in American popular culture, the show tended to trivialize the politics of oppression. As Esther Romeyn and Jack Kugelmass point out, the show often downplayed white/Indian conflict as acts of native protest were presented in the series as "purely symbolic and ludic." During Thanksgiving season in one episode, for instance, the Indian community takes the opportunity to pelt the town's white residents with tomatoes. Fleischman remains puzzled by the practice until Dave, the waiter, explains that it operates as a means for Indians to reclaim an otherwise unpleasant national holiday. "Could be worse," he tells Fleischman. Instead of tomatoes, they could have opted for "baseball bats, bicycle chains, tire irons." Aiming to reach a highly desired upper-income, well-educated TV viewership, the show often sidestepped overtly political concerns. As Romeyn and Kugelmass argue, in using Alaska as an offbeat, unconventional setting where difference prevails and all forms of oddities are forgiven, *Northern Exposure* ultimately revealed the "limitations of its redemptive vision," as the show's vision largely proved to be the "dreamscape of an audience of young, urban professionals."[9]

Oil-Age Alaskans

Even as one might concede to these problems, *Northern Exposure* nevertheless did much to expand on the roles Indians traditionally have played in the Western, especially in the character of Ed. In particular, the show offered a much-needed break with popular stereotypes of Alaska Natives as a people somehow frozen in time, isolated in their remote corner of the world where the forces of modernity have yet to arrive. With the discovery of oil in Prudhoe Bay in 1968, the subsequent settlement of land rights with the Alaska Native Claims Settlement Act (ANCSA), and the flood of oil money throughout the state, anxieties abounded across the nation concerning the future of the nation's Last Frontier. Regarded by many Euro-Americans as important links to the region's wild, untrammeled, pre-oil past, Alaska Natives soon became the objects of much of this displaced anxiety.

Nature writer John McPhee, for instance, who otherwise tries to dismantle mythic images of a wilderness Alaska in his book *Coming into the Country*, addresses in a less than complimentary manner what he refers to as the new class of "business-smart" natives. The new Alaska Natives unsettle McPhee primarily because of their increased mobility and access to the outside world. Traveling throughout the Lower 48, these once-isolated figures now successfully negotiate high-powered deals arising out of the establishment of native corporations in the wake of ANCSA. McPhee laments to readers that "in the Alaskan lexicon, a new synonym for 'native' was stockholder" (152). Discussing one particularly successful businessman, Willie Hensley, McPhee describes him as a "Brooks Brother native," a member of a new class of Eskimo who has seen and experienced a much wider segment of the world than his more isolated predecessors ever had. No longer confined geographically to a small corner of the globe, the new Alaska Natives occupy a much larger world. Such changes disturb McPhee, as they signal to him the profound transformations that capitalist development and modernization have brought to the nation's Last Frontier and its indigenous inhabitants. Gone are the dogsleds as a primary form of transportation. They are now replaced by snowmobiles. Fur parkas experience a similar fate as tailored suits take their place to become the new dress code for the corporate native.[10]

Journalist and travel writer Joe McGinniss offers an even more dire take on the new Alaska in his best-selling book, *Going to Extremes*, a dramatic account of developments that took place across the state in the post-pipeline era. Throughout his study, McGinniss describes his travels across the North, documenting the economic and social changes that have transformed Alaska. He meets one Eskimo, Al Cook, whom he initially describes with approval as a "survivor," someone who has remained impervious "to the assaults of time and progress upon the sacred traditions of his people." Later, however, McGinniss recounts a visit to his home in which Cook's eight-year-old son runs into the room where the two men are talking and breathlessly announces the beginning of their favorite television show. McGinniss captures the father's response: " 'Oh boy.' ... 'Hurry up. ... Six Million Dollar Man on TV.' " Television, in this account, becomes a sign for all that is wrong with the new Alaska. Noting the changes that have taken place around him, McGinniss expresses dismay at what the arrival of TV means for the community, whose new life, he believes, now revolves around the tube. "The children watched 'Sesame Street' each afternoon," he complains. "Little Eskimo kids coming off the tundra and sitting three feet from the screen. Learning to count from one to ten in Spanish. Gone was the symbolism of the raven and the bear. The new gods were Big Bird, and Bert and Ernie."[11] Ultimately, McGinniss's comments here say less about Alaska Native life in the post-pipeline era than they do about his own desires for a stable, premodern primitive, a figure that supposedly embodies elements of the national past which he and other white Americans have somehow lost.

In this context, the treatment of Ed in *Northern Exposure* serves as a breath of fresh air, an important departure from stereotypes of a doomed and Vanishing Alaskan that circulated in the oil era. Although it was not always fully successful in reshaping understandings of the region and its native residents, the show did manage to open up new representational terrain. Lately, a number of other texts have emerged that likewise show an interest in dismantling myths of the Last Frontier. In her memoir, *Road Song*, for instance, Natalie Kusz recounts her girlhood in Alaska; revising the standard boy-and-his-dog stories that Jack London popularized in the early twentieth century, Kusz tells about her experiences being attacked by wild dogs in an incident that cost her an eye.[12] Her story operates as a cautionary tale about uncritically embracing fantasies of the wild, a lesson that is usually underplayed in mainstream wilderness narratives. Like Kusz, Alaska writer Sheila Nickerson also paints a picture of the state that departs from the usual fare in *Disappearance: A Map*. As a meditation on the dangers that wilderness adventurers often face in the region, Nickerson's book recounts her experiences living in the North, juxtaposing everyday moments with stories about people who have died in freak accidents in nature (or in an "Alaskan death," as it is frequently referred to across the state). In a similar way, Nancy Lord provides a different take on Alaska myths. In "Marks," a story from her collection, *Survival*, Lord rewrites the frontier dream from a female perspective, providing a fictionalized account of a series of crimes actually committed during the 1980s against several women who became the prey of a crazed male hunter interested in having the ultimate wilderness experience in Alaska.

Anthropologist and nature writer Richard Nelson, who lived with the Koyukon Athabascans for many years, has also struggled with the legacy of the Last Frontier, especially as it emerges in the tradition of the nature essay. As he recounts in his book, *The Island Within*, the Koyukon offer a strikingly different view of nature from what he has learned in his culture.[13] They believe, for instance, that the forest trees are not inanimate objects but actually feel, hear, and sense one's presence among them. "There is no emptiness in the forest, no unwatched solitude, no wilderness where a person moves outside moral judgment and law" (13). The ideal of the solitary individual seeking solace in an uninhabited nature makes little sense for the Koyukon in Nelson's account. As such, his writings aim to dismantle the nature/culture dualisms that shape Euro-American responses to the landscape but which have not served nature writers particularly well throughout literary history.

Nelson also seeks to rethink geographical hierarchies that often structure nature writing, hierarchies that suggest certain regions are intrinsically more valuable or more worthy than others. He writes, "Although the island has taken on great significance for me, it's no more inherently beautiful or meaningful than any other place on earth. What makes a place special is the way it buries itself inside the heart, not whether it's flat or rugged, rich or austere, wet or arid, gentle or harsh, warm or cold, wild or tame" (xii). For Nelson, every place can be "elevated by the love and respect shown toward it" (xii). Aiming to imagine a different Alaska and a different way of writing nature, he reconsiders dominant landscape conventions, rethinking as well the social relations that are fostered by these traditions. The act of writing itself thus becomes subject to reconsideration. Noting how a common trope in the genre—the practice of naming geographical features on a map—functions within the larger colonial project of ownership, he imagines another way of writing. "I wish these names could be quietly abandoned, filed away as remnants of an archaic, human-centered attitude toward nature. For replacements, I would choose names given by Native American people, reflecting generations of physical and spiritual intimacy with the animals. Failing that, I would search for names among farmers and backwoods people, drawing on their common-sense observations and the poetry of vernacular speech. As a last resort, I would ask the biologists and birdwatchers, hoping for better luck this time." Nelson recognizes how the act of naming often functions to secure larger projects of domination and control; in its place, he suggests that each of us call the places we know by our own names with the hope that eventually they will die with us even as the land endures (134, 218).

If the physical maps that have inscribed Alaska are in need of revision, then so are the mental maps that shape the region. Nelson struggles in particular with images of Alaska as a pristine region, at times fighting against this depiction and at other times giving in to it. He refers to the area he visits, for instance, as an untouched wilderness even as numerous signs indicating otherwise are made present to him. During a stroll around the island at one point, Nelson encounters the ruins of a World War II bunker, its construction slowly being reclaimed by the earth. If lingering signs of the military's presence in the region are difficult to ignore, then the same proves to be true of the evidence left by the destructive logging and fishing industries. Nelson notes the effects of the timber industry on the island, whose system of roads leads to an area of logged-out hills that strongly resemble "a war zone." Later, he looks at the site of a clear cut, where for thirty or forty acres, not a living tree above sapling size remains. "I stand to see the whole forest of stumps. It looks like an enormous graveyard, covered with weathered markers made from the remains of its own dead" (38, 54).

Eventually, Nelson finds he must acknowledge the ways this beloved region is not an isolated or secluded space but a place already shaped by the forces of global capitalism. "The clearcut valley rumbled like an industrial city through a full decade of summers, as the island's living flesh was stripped away. Tugs pulled great rafts of logs from Deadfall Bay, through tide-slick channels toward the mill, where they were ground into pulp and slurried aboard ships bound for Japan. Within a few months, the tree that took four centuries to grow was transformed into newspapers, read by commuters on afternoon trains, and then tossed away" (56). Nelson is careful not to merely scapegoat Japanese industries or lay blame on populations located far from the forested valleys that are being logged. Instead, he recognizes these problems as an aspect of his own history and thus ponders his own connection to the complexities of development and the growth of world markets. Nelson writes candidly about his own links to these larger projects:

> I hold few convictions so deeply as my belief that a profound trans-gression was committed here, by devastating an entire forest rather than taking from it selectively and in moderation. Yet whatever judgment I might make against those who cut it down I must also make against myself. I belong to the same nation, speak the same language, vote in the same elections, share many of the same values, avail myself of the same technology, and owe much of my existence to the same vast system of global exchange. There is no refuge in blaming only the loggers or their industry or the government that consigned this forest to them. The entire society—one in which I take active membership—holds responsibility for laying this valley bare. (56–57)

Here Nelson is careful to note the ironies of his own position as a nature writer. As an individual who has benefited from numerous developments which have resulted, among other things, in the logging of Alaska's backcountry, he finds himself intimately tied to the changes he witnesses in the state, as are all readers of his book. The same process that brings the Japanese to Alaska in search of herring eggs, whose supply near home has been overfished, is the same process that brings anglers from the Lower 48 north to Alaska in search of re-sources that are in short supply elsewhere.

For the most part, Nelson's contributions to Alaska nature writing help redirect the genre in important ways. There are moments, however, when he appears unable to move outside some of the more troubling conventions of the form. Nelson understands, for instance, the interest that people might have in locating secluded, undiscovered land, and thus employs fictitious names throughout his book in an effort to respect the island's right to privacy and to "leave intact for others the privilege of discovering a place on earth" (xii). Although on one level the gesture appears generous to the land and his fellow nature enthusiasts, on another level it contributes to the colonial rhetoric often adopted by the environmentalist bent on becoming modern-day Robinson Crusoes or American Adams. At one point he explains, "Through my whole adult life, I've sought to experience unaltered, unbridled nature in wild places such as this. I've focused my work around it, chosen my home because of it, given up economic assurance to pursue it, made it a centering point of my existence. Sitting above this darkened beach, at the edge of the northern ocean, in the place that has so possessed me, I could scarcely ask for more" (93). At another point he confesses, "Sometimes I feel like a survivor from that age, a figure on a faded tintype, standing in a long-vanished pristine world. I read my scrawled island notebooks as if they've been discovered in someone's attic, recollections of a lost way of life. It creates a strange feeling of self-envy and romance, and makes me live this miracle all the more intensely" (58). Although Nelson tries to avoid scripting Alaska as an exceptional terrain set off from the rest of the country, a new world that can redeem a corrupted America, at times he is unable to get outside these narrative conventions. His writing thus becomes a testament to how strong these understandings remain in the dominant national imagination.

Critic David Mazel addresses mainstream environmental "structures of feeling" in ways that are useful to my discussion here. As he puts it, the concept of wilderness poses a problem for environmentalism in that it typically "misnames an anxiety as a geography."[14] As an imagined fantasy space, wilderness in recent years has undergone a great deal of critical scrutiny and is beginning to be understood as an ideological rather than a biological entity. For many critics, the meanings assigned to wilderness often signal the presence of larger cultural concerns—issues that have long shaped responses to places like Alaska. Carl Talbot considers these responses as symptoms of late industrial capitalism; for him, "the idea of wilderness is more than a 'matter of perception': it plays a vital role in the cultural logic of capitalism. As a sanctuary away from the degrading consequences of capitalist work relations, the idea of wilderness reduces nature to a 'leisure-time concept.' " The idea of wilderness, Talbot explains, operates as part of our myth of nature, concealing the facts of its production "by offering false compensation in the sphere of leisure consumption."[15]

Other critics have likewise remarked on the ways wilderness anxieties fuel problems of their own. Richard White, for instance, has recently commented on the strange form of environmental fundamentalism that has taken over many of the landscape debates in the region. White argues that nature advocates in the American West now often find themselves contending with a particular "fundamentalist strain of environmentalism" which conjures up images of a balanced, self-controlled nature, a pure, unsullied realm that is somehow independent of human beings and that should be forever shielded from use.[16] This kind of thinking, he suggests, is fundamentalist precisely because its understanding of a certain form of nature is rigid and misleading as well as strangely literalist with regard to nature. Like other strands of fundamentalist thinking, this type of nature advocacy operates within a closed circuit that prevents the entry of outside critique or debate. Ramachandra Guha also contributes to a rethinking of wilderness, addressing in particular the problems of wilderness preservation from a Third World perspective. More recently, he has focused on what he calls the "siege-like mentality of the wilderness lover," whose defense of the wild often adopts a militaristic demeanor.[17] Imported to other nations by the West's new green missionaries, the wilderness idea has often had negative consequences for non-Western populations. Guha further argues that the elevation of wilderness over other environmental concerns operates as a kind of "ecology of affluence," an observation that functions as a central argument motivating the growing environmental justice movement.[18]

But isn't all this talk about rethinking wilderness ideologies and recasting the Last Frontier a dangerous project, especially in the current political and economic climate where most Alaska lawmakers are given to being "green" only in a certain sense of the term and in fact would be only too happy to have more arguments about stepping up industrial development across the state? Isn't it perilous to question constructions of wilderness at a time when even the Arctic National Wildlife Refuge (ANWR) faces a dire future under George W. Bush's administration?[19] Are critiques of wilderness, in other words, too easily appropriated by those who lack a commitment to environmentalism? William Cronon addresses these concerns, directing us to the ways wilderness advocacy has often emerged in national life as a morally charged project. The promise of the wild, he argues, always touches on national myths of origin, particularly the myth of Eden, which "describes a perfect landscape, a place so benign and beautiful and good that the imperative to preserve or restore it could be questioned only by those who ally themselves with evil."[20] Poet Gary Snyder has made just this argument, contending that those individuals who address the constructedness of wilderness or nature must be regarded with extreme suspicion. For him, true environmentalists should always be wary of people who critique the givenness of wilderness, who see nature as a constantly changing element, and who find it difficult to insist on a nature/culture dualism. Such groups—which include a wide range of individuals from environmental justice activists to postmodern theorists—are indistinguishable in his view from the long discredited Wise-Use movement with its pseudo-environmental rhetoric masking a prodevelopment agenda.[21] Snyder himself admits that pristine nature probably does not exist anywhere in North America, yet goes on to advocate for something very close to it in what he calls a "renewable virginity."[22] His term carries enough baggage of its own that it needs little further comment here. Let it suffice to say that such a "renewable virginity" is not a gender-neutral concept and would be little valued from a feminist ecological perspective.

Unlike Snyder, other environmental critics refuse to concede that arguments about the constructedness of nature are necessarily damaging to the larger movement. Mazel, for example, argues that a myth of "pure nature" has often worked the other way, in fact, leading environmentalism "to undermine its most progressive aims by obscuring and enabling the economic, political, and historical relationships at the root of both environmental destruction and human oppression."[23] An environmentalism that questions the basic premises of wilderness or the givenness of nature even in an era of environmental decline need not be a destructive force for a progressive movement. As Mazel points out, wilderness might serve as a useful fiction in certain moments, employed as a kind of strategic essentialism; as a fiction, however, the term should still remain open to revision and rethinking. For Mazel, the larger environmental problem involves more than merely preserving a particular form of nature. Instead, it requires that we take care to not preserve "objectionable elements" of the larger social order sustaining that vision of nature.[24] Peter van Wyck concurs with this argument, contending that "an ecology without a critical practice can never be deep."[25] In terms of wilderness protection, numerous critics have indicated that the advocacy which has dominated certain forms of environmental activism in the United States is in fact in desperate need of a critical practice.

In the case of oil politics, wilderness, and Alaska, debates about protecting ANWR from development often overlook the bigger picture for oil companies. The strategy that the oil industry seems to be adopting today involves exploiting events such as the current West Coast electricity crisis in order to achieve a larger objective, which includes access to the Arctic National Wildlife Refuge as well as tax breaks for oil and gas extraction that could be worth billions of dollars. Even though oil provides only 3 percent of the source fuel used to generate electricity nationwide and only 1 percent in California, the recent energy crisis has emerged as a politically useful argument for oil companies. Furthermore, the big picture for the oil industry is not about opening ANWR to development but, rather, opening sites along the coast of Alaska, in the Gulf of Mexico, and in the Rocky Mountain West. Today the Department of Interior is looking at hundreds of lease applications that piled up during the Clinton years, which call for new offshore drilling that could bring oil development to the Beaufort Sea and the Copper River Delta (home of one of the greatest salmon runs in the world), Cook Inlet, Bristol Bay, and the Chukchi Sea near Point Hope—in other words, almost the whole coast of Alaska.

By understanding the oil industry's larger aims, we can better assess the strategic oversights on the part of the big national environmental groups who seem to be focusing their efforts almost entirely on ANWR, conceding without much struggle the rest of the coast of Alaska as well as other sites in North America. Meanwhile, local groups, including the Gwich'in, who are trying to protect ANWR and the National Petroleum Reserve west of Prudhoe Bay, and the Iñupiat Eskimos, who are fighting to defend their whale hunting grounds against development in the Beaufort Sea, understand quite well the real stakes involved. With ANWR figuring as a strategic symbol and rallying call for wilderness advocates across the United States, other sites in the state receive less attention and appear to be forfeited in the struggle to preserve select regions. Deborah Williams, executive director of the

Alaska Conservation Foundation, appeared on CBS's *Sixty Minutes* in the winter of 2001, where she addressed the fate of ANWR. While she vowed that not one oil rig would ever be built on the refuge, Williams explained at another point to the *New York Times* that she would nevertheless support oil drilling in the National Petroleum Reserve, a move that ultimately places her organization and its environmental efforts in direct conflict with many Alaska Natives living in and around the region. As this instance makes clear, the outcome of thinking in terms of wilderness is often an oversight that allows for the loss of what is not considered "ideal" nature. In practicing this kind of narrowly defined politics, national wilderness groups often end up relinquishing other areas with dire consequences to the people and wildlife who inhabit those lands.[26]

Environmental cultural studies—a field of knowledge influenced by a diverse constituency including cultural geographers, art historians, political scientists, literary critics, cultural historians, film and television scholars, postcolonial theorists, and even scientists—acknowledges that nature is a product of knowledge, something that is not just out there to be found but that is created within particular social contexts and for particular purposes. The interdisciplinarity of environmental cultural studies allows us to see the various interconnections of social phenomena that occur across forms of knowledge, an interconnection that is vital to the field of environmentalism as well. Given its genealogy in a variety of disciplines, environmental cultural studies is not beholden to the authority of any one discipline, even the authority associated with modern science and its stance of objectivity, a type of knowledge often employed by conservative politicians who deal with ecological matters only when the "best science" is made available to them (meaning the science that agrees with their political ideology). Rather, what environmental cultural studies recognizes is the human element of nature, the social histories that shape ideas about the environment, and the ways constructions of nature often contribute to larger social formations. As I have sought to indicate in the case of Alaska, much is at stake in ideas about nature, in ideas about the wilderness and ideas about the Last Frontier. To leave unacknowledged the problems associated with these concepts, I would argue, is a more dangerous project in the long run. The political and social dimensions of our concepts of nature and wilderness must be acknowledged and cataloged; if we fail to do so, we will remain firmly within the clutches of a way of thinking that has proven to be an environmental dead end. It remains crucial, then, that we continue to examine the ways Alaska's position as a state of nature is not a self-evident fact, but a deeply cultural phenomenon, a project that has been intimately linked to the development of American national identity and to larger processes of nation formation.

NOTES

Introduction

1. For this discussion of the economic development of Prince William Sound, see Wooley, "Alutiiq Culture," 140–41. Lethcoe and Lethcoe address the human presence in Prince William Sound from the Eskimos seeking better hunting grounds to the English and Spanish explorers searching for a Northwest Passage, and from the Russian adventurers hunting for sea otter pelts to the fox farmers, gold seekers, and Filipino cannery workers who all sought livelihoods in the region. For more on this history, see Lethcoe and Lethcoe, *History of Prince William Sound*, 1.

2. Wilson, *The Culture of Nature*, 286–87. For a discussion of the oil industry's routine violations of safety standards, see Smith, *Media and Apocalypse*, 77–114.

3. Daley, " 'Sad Is Too Mild a Word.' " For a general discussion of the ways race factors into environmental reporting, see Davis, *Ecology of Fear*, 51–52. For an overview of the nature/culture division in contemporary U.S. culture, see Price, *Flight Maps*, xv–xxii.

4. Alaska has figured as an object of fascination in the national imaginary in ways that seem to cut across divisions of race, gender, and history. The uses of Alaska as a sign of promise and possibility appear in a variety of literary texts throughout the twentieth century. Okanagon writer Mourning Dove, for instance, sends the main character's father off to the Alaska gold rush at the beginning of her 1927 novel, *Cogewea, the Half-Blood*. Meridel Le Sueur's working-class narrative, *The Girl*, which she began writing in 1939 and finished forty years later, figures Alaska as a mythical place of refuge for the novel's struggling workers. Arthur Miller's Pulitzer Prize–winning play, *Death of a Salesman* (1949), also features a reference to Alaska as a site of possibility. Uncle Ben, who has several properties he's looking at in the region, tells the family that "opportunity is tremendous" throughout Alaska. Toward the end of *Lolita* (1955), Vladimir Nabokov's love affair with the English language and the American landscape, Humbert Humbert receives a letter from Lolita telling him that she and her husband Dick plan to move to Juneau, Alaska, where he has been promised a "big job" in his own "specialized corner of the mechanical field" (268). In this allegory about America, the youthful nation, Nabokov demonstrates how the national dream has been forsaken by midcentury as both Lolita and the Last Frontier are trashed by the story's end. In Larry McMurtry's antiwestern novel, *The Last Picture Show* (1966), the burned-out, middle-aged Coach Popper envisions Alaska as an escape from his dead-end life in the small Texan town he calls home. Norman Mailer's *Why Are We in Vietnam?* (1967) also features Texans who long for bigger adventures, chronicling the experiences of a father and son who make their way to

Alaska's Brooks Range in order to engage in a hunting expedition where they test their manhood against the wilds of the Far North. In a similar way, James Dickey's 1970 novel *Deliverance* features characters who use Alaska as their standard of wildness, comparing the relatively undeveloped and unsullied backwoods of Georgia to the possibilities of the Last Frontier. The list goes on.

5. For more on how the nation's desire for renewal structures Euro-American ideas about the landscape, see Bercovich, *The American Jeremiad*.

6. For more on the economic ideas associated with the frontier, see Slotkin's *Fatal Environment*, 39–40, 531, and *Gunfighter Nation*, 18, 658.

7. Wilson, *The Culture of Nature*, 12.

8. My observations about the *Exxon Valdez* crisis build on arguments Raymond Williams has made about the "country and the city." As he explains, the anxious outcries concerning the destruction of the countryside do not solely concern threats to that particular landscape, but are instead ways of responding to larger historical developments. For further elaboration of these ideas, see Williams, *The Country and the City*.

9. Renato Rosaldo describes this yearning as "imperialist nostalgia" in *Culture and Truth*, 69.

10. As historians have argued, the closing of the frontier in the 1890s should be regarded as a myth. They have indicated in fact that the number of new homestead entries registered after 1900 was actually larger than the number of those registered during the previous three centuries. See, for instance, LaFeber, *The New Empire*, 64.

11. LaFeber, *The New Empire*, 25–26, 407–17. For other studies of American imperialist activities during the late nineteenth century, see Plesur, *America's Outward Thrust*; and Van Alstyne, *The Rising American Empire*.

12. Crapol and Schonberger, "Shift to Global Expansion," 136.

13. Sherwood, *Exploration of Alaska*, 4.

14. Alaska was not the first noncontiguous state to be admitted into the United States. California was also not contiguous to another state when it was admitted in 1850. For further discussion of this point, see Whitehead, "Noncontiguous Wests: Alaska and Hawai'i," 316.

15. LaFeber, *The New Empire*, 27–28.

16. The purchase of Alaska actually prompted quick global responses. According to Crapol and Schonberger, Canadians viewed the transfer of Alaska from Russia as a threatening act, the first indication that the United States was serious about taking over the northern half of the continent. Thus, in March 1867, the Dominion of Canada was formed, becoming official in the same month that Seward signed the treaty with Russia. For further discussion of these events, see Crapol and Schonberger, "Shift to Global Expansion," 132.

17. My discussion of the textuality of maps borrows from Harley, "Maps, Knowledge, Power," 229.

18. Sherwood, *Exploration of Alaska*, 37–38. In some ways, these depictions are not unique to Alaska, but operate as a trope developed by Western powers in the era of late imperialism. For a discussion of the ways various regions

were constituted as testing grounds for the West, see Plesur's analysis of U.S. imperialist rhetoric about the Congo in *America's Outward Thrust*, 144–56; and Bloom's examination of the "passion for blankness" and North Pole explorations in *Gender on Ice*, 1–15.

19. Sherwood, *Exploration of Alaska*, 16.

20. Samuel P. Hays points out that while the Progressive-era conservation movement is often cast as a moral struggle over the environment, it was not a broad popular outcry as many people would like to believe; instead, he argues that the movement stemmed from the realms of science and technology. Managed by professional experts, the conservation movement was motivated by the desire for rational planning that would encourage efficient development and use of natural resources in the interests of industry. See Hays, *Conservation and the Gospel of Efficiency*, 2.

21. Ibid., 189.

22. In arguing this point, I part ways with Samuel P. Hays and Barbara D. Hays, who contend that the two movements diverged as the resources considered important as commodities for conservationists became valued for their aesthetic uses by preservationists. See Hays with Hays, *Beauty, Health and Permanence*, 1–12.

23. See, for instance, Burnham, *Indian Country, God's Country*; Keller and Turek, *American Indians and National Parks*; Spence, *Dispossessing the Wilderness*; Schroeder and Neumann, "Manifest Ecological Destinies"; and Shiva, "The Greening of the Global Outreach."

24. Webb, *Yukon*, 307.

25. Goetzmann and Sloan, *Looking Far North*, xiv.

26. Andrew Ross discusses how environmentalist rhetoric during the Persian Gulf War operated to negate the nation's acts of conquest, suggesting that media reports frequently presented the U.S. military, the nation's biggest peacetime polluter, in the role of "Captain Planet," battling the imperial forces of ecoterrorism. For further discussion, see Ross, *Chicago Gangster Theory of Life*, 165.

27. Wilson, *The Culture of Nature*, 49.

28. Ibid., 43.

29. Haraway, *How Like a Leaf*, 50, 54.

30. Richard Dyer discusses Hollywood's racial ideologies of female desirability that required Marilyn Monroe to be "the most unambiguously white you can get." For more on this point, see Dyer, *Heavenly Bodies*, 43.

31. McClure, *Late Imperial Romance*, 8–29.

32. Cook, *Voyages Round the World*, 350.

33. Vancouver, *Voyage of Discovery*, 1305–7. For more on early voyages and the responses they generated about the region, see Murray, *Republic of Rivers*.

34. Nathaniel Portlock, quoted in Lambe, "Introduction," in Vancouver, *Voyage of Discovery*, 16.

35. James Strange, quoted in ibid.

36. Thanks to Doug Buege for bringing this ad to my attention.

37. Thomas, "Stream of Tourists Becomes a Torrent." Timothy Egan also reports that eleven years after the spill, many Alaskan towns that have been transformed by the cruise-ship industry since the 1980s are now trying to limit visits, with some residents fearing that Alaska is in danger of becoming "Disneyfied" by the rapidly growing tourism industry. See Egan, "A Land So Wild, It's Getting Crowded."

38. Jones, "Alaska's Trouble on Oily Waters."

39. This anxiety also informs the meanings assigned to the contemporary American West; the threat that the region is now somehow in decline encourages tourists to "see it before it disappears."

40. Chaloupka and Cawley, "The Great Wild Hope," 11.

41. Several critics have also made this point; see, for instance, Schmitt, *Back to Nature*, xix.

42. Worster, *Under Western Skies*, 216.

43. Wilson, *The Culture of Nature*, 287.

44. Seager, *Earth Follies*, 81.

45. Luke, "Green Consumerism," 155–56.

46. Ibid., 157.

47. Winner, *The Whale and the Reactor*, 137.

48. For more on this discussion, see Ross, *Chicago Gangster Theory of Life*, 4–5.

49. Noble, *The End of American History*, 6, 19.

50. Ibid., 19.

51. Kaplan, " 'Left Alone with America': The Absence of Empire in the Study of American Culture," 12–13. Edward Said points out, however, that ideologies of exceptionalism are not exceptional to the United States, as Western imperial powers have themselves frequently produced similar justifications in their drive for global dominance. For further discussion of this point, see Said, *Culture and Imperialism*, xxiii.

52. Kaplan addresses this problematic, explaining that "imperialism has been simultaneously formative and disavowed in the foundational discourses of American studies," in " 'Left Alone with America,' " 5.

53. Heyne, "The Lasting Frontier," 5, 15.

54. Foucault, "Questions on Geography," 69.

55. For further discussion of the racial history surrounding the concept of wilderness in U.S. history, see Darnovsky, "Stories Less Told," 15–16; and Gottlieb, *Forcing the Spring*, 4–6.

56. I wish to thank Joan Jensen for reminding me of this argument.

57. Samuel P. Hays and Barbara D. Hays, for instance, point out that radioactive iodine from the atomic explosions at Hiroshima and Nagasaki were found in lichen located in remote parts of Alaska and Lapland. In addition, lead from distant urban centers has been found in the relatively untouched regions of Greenland and the Antarctic. As they argue, because toxic elements migrate indiscriminately through air, water, and land, we cannot assume that

some regions remain protected from events happening elsewhere. For more on these points, see Hays with Hays, *Beauty, Health, and Permanence*, 172–73.

58. Gottlieb, *Forcing the Spring*, 19. He also expands traditional definitions of environmental advocacy; although he places the movement's origins in the rural West, he is also interested in recognizing the struggles for occupational and community health advocated by urban reformers outside the West, such as Jane Addams and Alice Hamilton. For more on this point, see Gottlieb, *Forcing the Spring*, 47–80.

59. Williams, "Ideas of Nature," 67, 70. For further elaboration on the uses of the term "nature," see Williams, *Keywords*, 184–89.

60. Miller, *Nature's Nation*, 11, 198.

61. Ibid., 199–203.

62. Ibid., 207.

63. Harvey, *Justice, Nature, and the Geography of Difference*, 171.

Chapter One

1. Krakauer, *Into the Wild*, 174. R. W. B. Lewis argues that the myth of the New World introduced a new heroic figure, the American Adam, who was "emancipated from history, happily bereft of ancestry, untouched and undefiled by the usual inheritances of family and race; an individual standing alone, self-reliant and self-propelling, ready to confront whatever awaited him with the aid of his own unique and inherent resources." For further discussion, see Lewis, *The American Adam*, 5.

2. Krakauer, *Into the Wild*, 72.

3. Heyne, "The Lasting Frontier," 9.

4. Scott Slovic addresses the ways "the enthusiasm of the newcomer" shapes contemporary environmental politics in "Occupancy and Authenticity in Western Environmental Literature."

5. While a forty-year gap exists between the publications of Marshall and McPhee, I do not mean to suggest that there weren't other writers narrating Alaska in this period. I focus on these figures because they produced the most popular narratives about the region; their texts are still more widely referenced than other writers' work about Alaska.

6. Hochman, *Green Cultural Studies*, 8–9.

7. Quoted in Winkler, "Inventing a New Field." For some critics, the concern is also that ecocriticism seems to be attracting academic luddites, scholars who are largely fearful of contemporary critical theory. For a particularly astute analysis of the place of literary theory in environmental studies, see also Phillips, "Ecocriticism."

8. Haraway, *Simians, Cyborgs, and Women*, 189.

9. Roorda, *Dramas of Solitude*, xiii.

10. Ibid., 5.

11. Unfortunately, this line of thinking often informs ecocriticism and is

present even in Jhan Hochman's otherwise sympathetic treatment of green cultural studies. Because he describes postmodern theories of nature as a social construction as "seeing ourselves as gods," he appears unable to move beyond a transcendent understanding of nature. See Hochman, *Green Cultural Studies*, 11.

12. Muir, *Travels in Alaska*, 11–12. Further references will be cited parenthetically in the text.

13. Buell, *The Environmental Imagination*, 68–75.

14. Nash, "Tourism, Parks and the Wilderness Idea," 5–6.

15. Muir, *Letters from Alaska*, 30–31. Further references will appear parenthetically in the text.

16. Young, *Alaska Days with John Muir*, 64. Further references will appear parenthetically in the text.

17. Roderick Nash points out that Charles E. S. Wood entered the same fjord with Indian guides two years earlier. Wood, however, did not write about his trip until later, which enabled Muir to consider himself its explorer. In 1899, the Harriman Expedition traveled to Glacier Bay; its members also found a glacier that had just receded enough to allow passage into the newly created fjord. Muir was aboard this expedition, as was the geographer Henry Gannett, who wrote an account of discovery much like Muir had nearly twenty years before. For further discussion of these events, see Nash, "Tourism, Parks, and the Wilderness Idea," 8.

18. Udall, *The Quiet Crisis*, 144.

19. For a discussion of nature writing as a specimen and trophy of the exotic, see Stewart, *On Longing*, 147. Aldo Leopold also addresses the problem of wilderness excursions and trophy hunting in *A Sand County Almanac*, 176. Leopold, however, doesn't provide a detailed examination of the colonizing gestures involved in trophy hunting and its effects on the region's Indian populations.

20. My discussion of nature preservation and the colonial museum borrows from Root, *Cannibal Culture*, 21. Muir also expresses pleasure at the raw materials Alaska offers during the 1881 Corwin Expedition which took formal possession of Wrangell Land, claiming it for the United States. He remarks that a "land more severely solitary could hardly be found anywhere on the face of the globe," in *The Cruise of the Corwin*, 134.

21. Jakle, *The Tourist*, 23, 42.

22. Nash, *Wilderness and the American Mind*, 161–81.

23. Worster, *Under Western Skies*, 14.

24. Muir, *Our National Parks*, 7.

25. Ibid.

26. Rodriguez, "Art, Tourism, and Race Relations in Taos," 156.

27. For a discussion of nature as a national monument, see Hays with Hays, *Beauty, Health, and Permanence*, 57; and Gottlieb, *Forcing the Spring*, 29.

28. Quoted in Goetzmann and Sloan, *Looking Far North*, 14.

29. Burroughs et al., *Alaska*, 69.

30. See Boime, *The Magisterial Gaze*, 3–8; and Mitchell, *Witnesses to a Vanishing America*, 31.

31. Muir, *Edward Henry Harriman*, 3–4.

32. For a more detailed discussion of the "greening" of American capitalism in the modern era, see Wilson, *The Culture of Nature*, 79–81.

33. Buzard, *The Beaten Path*, 88.

34. Muir's evasions about tourism connect him to a larger myth in American literary history concerning the idea of the "innocent eye." In his classic study, Tony Tanner addresses the ways dominant American literature, beginning in the nineteenth century, continually expresses a distrust for judgement and analysis. In its place, the naive or innocent eye becomes cultivated by writers, especially those interested in nature. The celebration of "passive wonder" in turn, enables figures such as Muir to imagine that they are somehow outside history, outside ideology, even outside power relations. See Tanner, *The Reign of Wonder*.

35. Frow, "Tourism and the Semiotics of Nostalgia," 146.

36. Buzard, *The Beaten Path*, 4.

37. Ibid, 8.

38. White, "Are You an Environmentalist or Do You Work for a Living?," 171–72.

39. Darnovsky, "Stories Less Told," 17, 26.

40. Ibid., 16–17.

41. For a discussion of the impact that wilderness advocacy and the creation of national parks had on native land tenure in Southeast Alaska, see Catton, *Inhabited Wilderness*, 9; Mitchell, *Sold American*, 164; and Staton, *A National Treasure or A Stolen Heritage*. See also the oral histories of Glacier Bay in James, "Glacier Bay History;" and Marvin, "Glacier Bay History." For a more optimistic view of the history of nature advocacy in Alaska, see Ross, *Environmental Conflict in Alaska*.

42. See Brown, "Feds Put Gull-Egg Gathering on Ice."

43. For further discussion of the overlapping histories of development and aesthetics, see Wilson, *The Culture of Nature*, 42–47.

44. Worster, *Under Western Skies*, 168–69.

45. Glover, *A Wilderness Original*, 5.

46. Gottlieb, *Forcing the Spring*, 15.

47. Ibid., 16.

48. Ibid., 17.

49. Ibid., 19.

50. Marshall, *Alaska Wilderness*, 97.

51. Ibid.

52. Ibid., 109.

53. Glover, *A Wilderness Original*, 4.

54. For a discussion that relates mountaineering essays to a similar project of unmapping, see Ellis, "A Geography of Vertical Margins."

55. Marshall, *Arctic Village*, 3.

56. Marshall, *Alaska Wilderness*, 1–2.

57. Ibid., 44.

58. Ibid., 89.

59. Ibid., 39.

60. Ibid., 40–41.

61. In many ways, Marshall's other narrative, *Arctic Village*, resembles Thoreau's account of his excursion to Walden Pond. Marshall, for instance, rewrites Thoreau's essay on "Economy," detailing his three years in the Far North with extensive charts and graphs that provide an account of how he spent his time and money during his stay in Wiseman.

62. Marshall, *Alaska Wilderness*, 3–4.

63. Ibid., 4.

64. Nerlich, *The Ideology of Adventure*, 80–81.

65. Ibid., 209.

66. The vast literature on the sociology of tourism and travel also makes this point. For a discussion of the ways travel serves to delineate class identities, see, for instance, Buzard, *The Beaten Track*; Jakle, *The Tourist*; Leed, *The Mind of the Traveler*; and Wilson, *The Culture of Nature*.

67. Glover, *A Wilderness Original*, 94–95.

68. See Schmitt, *Back to Nature*, 15; and Wilson, *The Culture of Nature*, 227.

69. Knobloch, *The Culture of Wilderness*, 19–20.

70. Leopold, "Preface," *Alaska Wilderness*, v.

71. Marshall, *Alaska Wilderness*, 46.

72. For more on tropes of emptiness in landscape descriptions, see Duncan and Duncan, "Ideology and Bliss."

73. Quoted in Marshall, "Introduction to the First Edition," *Alaska Wilderness*, xxxii–xxxiii.

74. Williams, *Problems in Materialism and Culture*, 77.

75. Marshall, *Alaska Wilderness*, 113.

76. McPhee, *Coming into the Country*, 126. Further references will be incorporated parenthetically in the text.

77. Birkland, *After Disaster*, 99.

78. Ibid., 86.

79. Ibid., 85.

80. Ibid., 104.

81. Lopez, *Arctic Dreams*, 355. Further references will be incorporated parenthetically in the text.

82. Kuhls, *Beyond Sovereign Territory*, ix–xiv.

83. Ibid., x.

84. Ibid., 127.

85. Seager, *Earth Follies*, 147.

86. For further discussion of the interconnected histories of environmentalism and Western colonialism, see Cosgrove, "Habitable Earth"; Grove, *Green Imperialism*; Guha, "Radical American Environmentalism and Wilderness Preservation"; and Tsing, "Transitions as Translations."

87. Tsing, "Transitions as Translations," 258.

88. Ibid., 259.

89. "Alaska Native" is the term of choice used to collectively address the state's indigenous peoples. Athabascan Indians live primarily in the interior of Alaska. Tlingit, Tsimshian, and Haida Indians live in Southeast Alaska. Yup'ik Eskimos reside primarily in Southwest Alaska, while Siberian Yup'iks live on St. Lawrence Island. Iñupiaq Eskimos live mostly in Northwest Alaska and on the North Slope, while Aleuts reside along the Aleutian Islands. For further discussion of the uses of the terms "Alaska Native" and "Alaska Eskimo," see Andrews and Creed, *Authentic Alaska*, xxvii; and Fienup-Riordan, *Freeze Frame*, xix.

90. O'Neill, *The Firecracker Boys*, 270.

91. Ibid., 26.

92. Ibid., 238.

93. Haines, "Foreword," in Muir, *Travels in Alaska*, x.

94. In arguing for a recognition of the invention of Alaska as a wilderness region, I am indebted to the ideas of Cronon in "The Trouble with Wilderness."

95. Silko, *Yellow Woman and a Beauty of the Spirit*, 94.

Chapter Two

1. Norris, "The Frontier Gone at Last," 1728–29.

2. Several critics have discussed the use of U.S. frontier rhetoric in depicting overseas expansion. See, for instance, Drinnon, *Facing West*; and Kaplan, "Romancing the Empire." Other U.S. writers who used the Far North for their frontier settings include Emerson Hough, Robert Service, Hamlin Garland, and Frank Norris. Interestingly, the first figure hailed as a "real Alaskan novelist" was not a male but a female writer, Barrett Willoughby. For a study of her literary accomplishments during the 1920s and 1930s, see Ferrell, *Barrett Willoughby*.

3. This struggle shares much in common with the project that McClure addresses in *Late Imperial Romance*.

4. Turner, *The Frontier in American History*, 37, 206.

5. Ibid., 293.

6. Many critics have argued this point; for representative discussions, see Maltby, "John Ford and the Indians"; and Wrobel, *The End of American Exceptionalism*.

7. Turner, *The Frontier in American History*, 296.

8. Bloom, *Gender on Ice*, 1–3.

9. Canadian historian Pierre Berton refers to Dawson as an "American city on Canadian soil" in the documentary film *City of Gold* (1957). For a history of the Klondike gold rush, see Berton, *The Klondike Fever*; and Porschild, *Gamblers and Dreamers*.

10. In constructing a conceptual framework for the study of northern

regions, Ken Coates suggests that a transnational approach might enable scholars to better understand such areas. Unlike London, however, Coates is sensitive to the ways political boundaries of the nation-state must also be recognized in studies of the North. For more on this point, see "The Rediscovery of the North," 15–43.

11. For further discussion of the modern "ecological subject," see Luke, "Green Consumerism," 156.

12. Cosgrove, "Habitable Earth," 36.

13. The literature on the turn-of-the-century crisis in Anglo-Saxon masculinity is vast. For a representative sample, see Cosgrove, "Habitable Earth," 34–36; Broun, "Foreword," in Truettner, *The West as America*, vii–ix; Peterson, "Jack London and the American Frontier"; Schmitt, *Back to Nature*; and Bederman, *Manliness and Civilization*.

14. Schmitt, *Back to Nature*, 23.

15. For more on the problems of "wilderness design," see Chaloupka and Cawley, "The Great Wild Hope," 5–6.

16. Many historians have noted the imperial politics of U.S. preservationism. For a representative discussion, see Darnovsky, "Stories Less Told," 15–16; and Gottlieb, *Forcing the Spring*, 27.

17. For a useful discussion of Roosevelt and the back-to-nature movement, see Gottlieb, *Forcing the Spring*, 213; and Peterson, "Jack London and the American Frontier," 138.

18. For more information about London as a writer of the Klondike, see Walker, *Jack London and the Klondike*.

19. The dime novels of the period also misrecognize the Klondike as U.S. terrain. The confusion is evident in the titles and plot summaries provided in Leithead, "Tales of Klondike Gold in Dime Novels." For a discussion of the divided and contested memory of Jack London among some Canadian residents, see Campbell, "Facing North."

20. London, "The God of His Fathers," *Complete Short Stories*, 1:382.

21. London, *Revolution and Other Essays*, 180.

22. For further discussion of the Western's origins in the British colonial adventure narrative, see Cawelti, *The Six-Gun Mystique Sequel*, 25; and Slotkin, *Gunfighter Nation*, 194.

23. Review of *The Son of the Wolf*, *The Beacon*, n.d., n.p. Collected in "The Scrapbooks," Jack London Papers, Huntington Library, San Marino, California.

24. Review of *The Son of the Wolf*, *The Wave*, May 12, 1900, n.a., n.p. Collected in ibid.

25. Review in *The Evening Transcript*, n.d, n.p. Collected in ibid.

26. "Alaskan Stories," *The Atlantic*, April 1900, n.a., n.p. Collected in ibid.

27. Quoted in Labor, "Jack London," 381.

28. Starr, *Americans and the California Dream*, 212.

29. London, *Smoke Bellew*, 1. Further references will appear parenthetically in the text.

30. London, *The Call of the Wild*, 21.

31. For a discussion of the ways this confusion informs Hollywood's production of Canada, see Berton, *Hollywood's Canada*. In a similar vein, Sherrill Grace refers to what she calls "literary Manifest Destiny," the process whereby U.S. critics claim Canadian writers as part of their own national tradition. For more on this idea, see Grace, "Comparing Mythologies: Ideas of West and North," 249.

32. London, "To the Man on Trail," *Complete Stories*, 1:157.

33. London, *Burning Daylight*, 38. Further references will appear parenthetically in the text.

34. According to Alfred Hornung, tropes of expansion play a central role in other London texts. In *John Barleycorn*, for instance, alcohol operates as a means of stimulating expansion; while under its influence, London feels that "all the world was mine." For more on this idea, see Hornung, "Evolution and Expansion in London's *The Road* and *John Barleycorn*."

35. For further discussion of the environmental impact of the stampede, see Mayer, *Klondike Women*; and Porschild, "The Environmental Impact of the Klondike Stampede."

36. London, *Daughter of the Snows*, 16. Further references will appear parenthetically in the text.

37. For a historical discussion of this environmental approach, see Hays, *Conservation and the Gospel of Efficiency*.

38. Limerick, *Something in the Soil*, 215.

39. Here London replaces frontier rhetoric with the discourse of urban wilderness. For more on how U.S. writers imagined the city as a new wilderness needing to be tamed, see Light, "Urban Wilderness."

40. Cook-Lynn, *Why I Can't Read Wallace Stegner and Other Essays*, 29.

41. Cosgrove, "Habitable Earth," 39.

42. My discussion of the emergence of Canadian northern iconography comes from Osborne, "The Iconography of Nationhood in Canadian Art." For an analysis of the distinctions between the "Canadian Northern" and the "American Western" and the ways they fit into different national narratives, see Grace, "Comparing Mythologies." For a useful overview of Canadian literary constructions of the "mystic North," see Atwood, *Strange Things*.

43. For a discussion of non-Canadian literary uses of the Far North, see Gross, "From American Western to Canadian Northern."

44. For a discussion of Burning Daylight as a Leatherstocking figure, see Watson, *The Novels of Jack London*, 170–71.

45. Quoted in Graebner, *Manifest Destiny*, 330.

46. Quoted in LaFeber, *The New Empire*, 28.

47. Many historians have addressed these events; see, for instance, Sherwood, *Exploration in Alaska*; Van Alstyne, *The Rising American Empire*; and Kiernan, *America, The New Imperialism*.

48. For more on U.S. expansion in the Canadian North, see Honderich, *Arctic Imperative: Is Canada Losing the North?*; Coates and Morrison, *The*

Alaska Highway during World War II; and Nielson, *Armed Forces on a Northern Frontier*.

49. See entry for "Rex Beach" in Buscombe, *The BFI Companion to the Western*, 65.

50. Beach, *The World in His Arms*, 84. Further references will be incorporated parenthetically in the text.

51. See Hutchinson, "The Cowboy in Short Fiction," 516.

52. Beach, *The Spoilers*, 1.

53. Beach, *The Winds of Chance*, 8. Further references will appear parenthetically in the text.

54. McPhee addresses this history in *Coming into the Country*, 227–28.

55. Robinson, *Having It Both Ways*.

56. Andrews, "Alaska and Its Gold Fields," 4.

57. Harris, *Alaska and the Klondike Gold Fields*, iii.

58. See Naske and Slotnick, *Alaska*, 87.

59. "Decision Criticised," 1.

60. See Coates, *Best Left as Indians*, xv, 35–41.

61. George Carmack wrote an interesting series of accounts concerning the strike. In the years following the Klondike stampede, he produced at least three versions of his story, each one more elaborate and embellished than the first. The first two accounts are collected in the Snow Papers, MS 38, Alaska State Library, Historical Collections, Juneau, Alaska. Another version, *My Experiences in the Klondike*, privately printed by Marguerite Carmack, 1993, may be found in the Beinecke Library at Yale University. For Native accounts of the Klondike strike, see Cruikshank, *Life Lived Like a Story*, 128–39. Cruikshank's account foregrounds Shaaw Tláa's role in particular. For more on Skookum Jim's role, see Rab Wilkie and the Skookum Jim Friendship Centre, *Skookum Jim*, copy in Alaska State Library, Historical Collections.

62. For an analysis of how myths of Canadian national identity also operate to erase the history of First Nations people, see Clarke, "White Like Canada."

63. Hepler, "Michigan's Forgotten Son"; and Curwood, *Son of the Forests*, 200–201.

64. Curwood, *The Alaskan*, 12. Further references will appear parenthetically in the text.

65. Quoted in Worster, *Under Western Skies*, 182.

66. For a discussion of how Molly Wood's revolutionary background figures into the logic of the novel, see Cawelti, *The Six-Gun Mystique Sequel*, 72–73. Incidentally, Wister's frontier romance also features a reference to a character who makes a journey to Alaska before returning to Wyoming. See Wister, *The Virginian*, 122.

67. White, *The Eastern Establishment and the Western Experience*, 181.

68. Ibid., 176–77.

69. For a discussion of ecology and the rhetoric of efficiency, see Hays, *Conservation and the Gospel of Efficiency*.

70. Doyle, *North of America*, 1, 150.

71. See McCarthy, "Zhirinovsky Upsets D.C.," 5; and Egan, "Alaskans Don't Want to Be Anyone's Siberia," 3.

Chapter Three

1. McClintock, *Imperial Leather*, 40–41. This discussion builds on Johannes Fabian's work in anthropology, specifically his studies of "chronopolitics" and the "denial of co-evalness" in Western culture. For further elaboration of these points, see Fabian, *Time and the Other*.

2. McClintock, *Imperial Leather*, 30.

3. For more on European fears about racial degeneracy, see ibid., 43.

4. See Nelson, "I Was a Bride of the Arctic"; Keen, "Woman in the Wilderness"; Hughes, "Honeymooning in Alaskan Wilds"; and Hendricks, "I Teach French in Alaska."

5. Scidmore, *Alaska*.

6. Catton makes this point in *Inhabited Wilderness*, 258. For more on Crisler's history, see Burg, "The Crislers—Brooks Range Naturalists," 9.

7. The Crislers also filmed sequences for other Disney nature films, including *The Olympic Elk* and *The Vanishing Prairie*. For further discussion of their Disney experiences, see Mitman, *Reel Nature*, 116–19.

8. Terry Tempest Williams describes Murie as the "spiritual grandmother" of the environmental movement in " 'You Have to Know How to Dance:' The Inspiration of Margaret Murie;" Debbie Carter refers to Murie as a "mother" to the modern environmental movement in "Film Frames Life of Conservationist."

9. Breton, *Women Pioneers for the Environment*, 259–60.

10. I borrow the term "strategy of return" from Roorda, who discusses this aspect of nature writing in *Dramas of Solitude*.

11. Crisler, *Arctic Wild*, 4. Further references to this book will be noted parenthetically in the text.

12. For a discussion of the changing views toward wildlife captivity in this period, see Norwood, *Made from This Earth*, 240–41. For a discussion of nature/culture boundary crossing, see Haraway, *Primate Visions*, 149.

13. See King, "The Audience in the Wilderness," 61–64.

14. For a discussion of the cultural work Disney films performed in this period, see Wilson, *The Culture of Nature*, 125; and King, "The Audience in the Wilderness," 61. For a discussion of the Disney Company's ambivalent position concerning the educational and entertainment value of their nature documentaries, see Bousé, *Wildlife Films*, 65.

15. Mitman, *Reel Nature*, 112–14.

16. Wilson, *The Culture of Nature*, 118; King, "The Audience in the Wilderness," 61–63.

17. De Roos, "The Magic Worlds of Walt Disney," 50.

18. Mitman, *Reel Nature*, 114.

19. Wilson, *The Culture of Nature*, 118–19.

20. Schickel, *The Disney Version*, 278.

21. See, for instance, Armbruster, "Creating the World We Must Save," 224; King, "The Audience in the Wilderness," 67; and Schickel, *The Disney Version*, 277–87.

22. Whitehead, "Alaska and Hawai'i," 195–97.

23. Wilson, *The Culture of Nature*, 122.

24. For further discussion of these events, see King, "The Audience in the Wilderness," 66; and Jackson, *Walt Disney*, 87.

25. Wilson, *The Culture of Nature*, 123.

26. See King, "The Audience in the Wilderness," 65. For a critique of Disney's rendition of nature in the animated films, see Murphy, " 'The Whole Wide World Was Scrubbed Clean.' "

27. Wilson, *The Culture of Nature*, 122.

28. Armbruster, "Creating the World We Must Save," 232.

29. See Burnham, *Captivity and Sentiment*, 1–9.

30. Crisler, *Captive Wild*, 40.

31. For more on ideologies of female exceptionalism and nation-building projects in Alaska, see Kollin, "The First White Women in the Last Frontier," 122.

32. Gaard, "Living Interconnections with Animals and Nature," 1.

33. Przybylowicz, "Toward a Feminist Cultural Criticism," 294.

34. Merchant, *Earthcare*, xi.

35. Kolodny, "Turning the Lens on 'The Panther Captivity,' " 331.

36. A new body of literature on the relationship between domesticity, social control, and territorial conquest has emerged in recent years. For a representative sample, see Armstrong, *Desire and Domestic Fiction*; George, "Homes in the Empire, Empires in the Home"; Georgi-Findlay, *The Frontiers of Women's Writing*; Kaplan, "Manifest Domesticity;" Kollin, "The First White Women in the Last Frontier;" and McClintock, *Imperial Leather*.

37. McClintock, *Imperial Leather*, 34.

38. Kaplan, "Manifest Domesticity," 584.

39. Ibid., 582.

40. Ibid., 582–83.

41. Ibid., 582.

42. Soper, *What Is Nature?*, 229.

43. Ibid., 228.

44. For a discussion of how popular culture mediated Cold War fears about America's atomic future, see Kupfer, "The Cold War West as Symbol and Myth," 169.

45. May, *Homeward Bound*, 14.

46. See Egan, "Everyone Is Always on Nature's Side;" and Martin, "Alaska's Dumb Dance with Wolves."

47. Berger, *About Looking*, 6.

48. Ibid., 9.

49. Ibid., 10.

50. Ibid., 13, 19

51. Catton, *Inhabited Wilderness*, 90.

52. For a discussion of the Crislers' work in Alaska, see Burg, "The Crislers—Brooks Range Naturalists."

53. See McClintock, *Imperial Leather*, 31–36; and Kaplan, "Manifest Domesticity," 584.

54. For more on how white settlers often portray Alaska Natives as bad ecologists, see Catton, *Inhabited Wilderness*, 180–81.

55. Haraway, *Primate Visions*, 7.

56. For a useful elaboration on Berger's argument, see Wolch and Emel, "Preface," *Animal Geographies*, xi, xvii.

57. Haraway, *Simians, Cyborgs, and Women*, 11.

58. See Baker, *Picturing the Beast*, 29; and Elder, Wolch, and Emel, "*Le Practique Sauvage*," 72–73.

59. For further discussion of these points, see Elder, Wolch, and Emel, "*Le Practique Sauvage*," 83.

60. Ibid, 80.

61. Breton, *Women Pioneers for the Environment*, 261.

62. "Biographical Sketch." Margaret Murie Collection. Alaska and Polar Regions Department, Rasmuson Library, University of Alaska, Fairbanks.

63. "Biographical Sketch"; Murie, *Two in the Far North*, 82. Further references will appear parenthetically in the text.

64. Breton, *Women Pioneers for the Environment*, 261; Jettmar, "A Meeting of Generations."

65. Carter, "Film Frames Life of Conservationist."

66. For accounts of the history of the modern environmental movement, see, for instance, Gottlieb, *Forcing the Spring*; and Hays with Hays, *Beauty, Health, and Permanence*.

67. The environmental justice movement, a coalition of grassroots organizers, has launched the most vocal critique of the mainstream environmental movement and its goals. For a representative sample of this countermovement, see the essays in Hofrichter, *Toxic Struggles*; Bullard, *Unequal Protection*; Westra and Wenz, *Faces of Environmental Racism*; and Camacho, *Environmental Injustices, Political Struggles*.

68. Murie, "Forum on Growth in Alaska," Margaret Murie Collection. Box 2, Series 2, Folder 22. Alaska and Polar Regions Department, Rasmuson Library, University of Alaska, Fairbanks.

69. Ibid.

70. *Hearing before the Merchant Marine and Fisheries Subcommittee of the Committee on Interstate and Foreign Commerce*, 59–60. Copy in the Margaret Murie Collection. Box 2, Series 2, Folder 31. Alaska and Polar Regions Department, Rasmuson Library, University of Alaska, Fairbanks.

71. Wilson describes Alaska as an "annex" to the West in *The Culture of Nature*, 120.

72. Murie, Letter of September 7, 1967, Margaret Murie Collection, Box 1,

Folder 8, Alaska and Polar Regions Department, Rasmuson Library, University of Alaska, Fairbanks.

73. Murie, "Forum on Growth in Alaska," 8–9.

74. Ibid., 11.

75. Ibid., 13, 17–18.

76. Van Wyck, *Primitives in the Wilderness*, 83.

77. For useful discussions of fetishism, see the essays in Apter and Pietz, *Fetishism as Cultural Discourse*.

78. White, "The Current Weirdness of the West," 13.

79. Cronon, "The Trouble with Wilderness," 80–81.

80. See Catton, *Inhabited Wilderness*, 81–82; and Staton, *A National Treasure or a Stolen Heritage*, 10.

81. For a discussion of the unfair treatment Alaska Natives faced under wildlife management laws, see Mitchell, *Sold American*, 192–251; for more on the conflicts between Alaska Natives and white environmentalists, see Catton, *Inhabited Wilderness*, 194–95.

82. Carrighar, Letter of April 24, 1956, Margaret Murie Collection, Box 2, Series 2, Folder 26, Alaska and Polar Regions Department, Rasmuson Library, University of Alaska, Fairbanks.

83. For a related discussion of how the illusion of being "alone in nature" operated among primatologists, see Haraway, *Primate Visions*, 154.

84. Catton, *Inhabited Wilderness*, 191–214. In that sense, Catton's argument counters Dave Foreman's claims that in Alaska, wilderness conservation has unproblematically allowed the inclusion of the region's indigenous people. For more on his argument, see Foreman, "Wilderness Areas for Real," 400–401.

85. Merchant, *Earthcare*, xv.

86. McClintock, *Imperial Leather*, 6.

Chapter Four

1. Hofrichter, "Introduction," *Toxic Struggles*, 1–10.

2. See Cronon, "The Trouble with Wilderness." For another critique of the uses of wilderness among mainstream environmentalists, see Dryzek's discussion of "green romanticism" in *The Politics of the Earth*, 155–71; see also Phillips's discussion of the place of nature in ecocriticism in "Ecocriticism, Literary Theory, and the Truth of Ecology."

3. See, for instance, the essays in Soulé and Lease, *Reinventing Nature?*

4. Cronon, "The Trouble with Wilderness: A Response," 54–55. For a discussion of race, gender, and the problems of wilderness, see White, "Black Women and Wilderness."

5. A word about terminology is in order here. As anthropologist Ann Fienup-Riordan explains, "In many parts of the Arctic, including northern Alaska, Eskimo peoples prefer to be referred to as Inuit. This preference

derives at least in part from the pejorative and derogatory connotations of 'Eskimo,' whose common etymology is 'eaters of raw flesh.' Actually the name comes from a Montagnais form meaning 'snowshoe-netter' (Goddard, cited in Damas 1984:6). . . . Yup'ik Eskimos, unlike their Canadian and Greenlandic neighbors, have not embraced the designation 'Inuit,' and they continue to refer to themselves as either 'Yupiit' or 'Yup'ik Eskimos.'" Many residents of Alaska thus use the terms "Eskimo" and "Alaska Eskimo" to refer to Inuit, Iñupiaq, and Yup'ik peoples in general and use the specific name when speaking of a particular group of people. See Fienup-Riordan, *Freeze Frame*, xix. For a discussion of the uses of the term "Alaska Native," see also Andrews and Creed, *Authentic Alaska*, xxvii.

6. LaDuke, *All Our Relations*, 131.

7. Schroeder and Neumann, "Manifest Ecological Destinies," 322.

8. Ibid., 321.

9. Seager, "Creating a Culture of Destruction," 59. See also Wolfley, "Ecological Risk Assessment and Management."

10. Price, *Flight Maps*, xvii–xviii.

11. Forbes, "Nature and Culture," 6.

12. Dauenhauer and Dauenhauer, "Preface," *Haa Kusteeyí, Our Culture*, xxxii.

13. Dauenhauer, *The Droning Shaman*, 25. Further references to this collection will be incorporated parenthetically into the text as *DS*.

14. LaDuke, *All Our Relations*, 149.

15. Dauenhauer and Dauenhauer, *Haa Kusteeyí, Our Culture*, 72.

16. Dauenhauer, *Life Woven with Song*, 61. Further references to this collection will be incorporated parenthetically into the text as *LW*.

17. In ways that connect with this poem, James Clifford writes tellingly about how descriptions of cultural Others often operate as a form of collecting, in "On Collecting Art and Culture," 141.

18. For more on the anti-Indian movement, see LaDuke, *All Our Relations*, 122.

19. Shiva, "The Greening of the Global Reach," 149.

20. Ibid., 151.

21. Hamilton, "Environmental Consequences," 69. Richard Kerridge describes this form of environmentalism as a "pretext for the wealthy to deny others the right to catch up." For more on this discussion, see Kerridge, 'Small Rooms and the Ecosystem."

22. Van Wyck, *Primitives in the Wild*, 2.

23. Shiva, "The Greening of the Global Reach," 152.

24. Chow, *Writing Diaspora*, 3–4.

25. See Gallagher, *American Ground Zero*; and Davis, "Dead West."

26. For biographical information on Robert Davis, see Bruchac, *Raven Tells Stories*, 48–49; and Niatom, ed., *Harper's Anthology of Twentieth-Century Native American Poetry*, 363.

27. Davis, *Soul Catcher*, 27. Further references to this book will be included parenthetically in the text.

28. Muir, *Travels in Alaska*, 245.

29. Ruppert, " 'Listen for Sounds,' " 86.

30. Ibid., 89.

31. For a fascinating discussion of the Presbyterian missionary Sheldon Jackson and his own history as a major collector of tribal arts in Alaska, see Lee, "Zest or Zeal?"

32. See, for instance, Bullard, "Environmental Blackmail in Minority Communities."

33. See Dauenhauer and Dauenhauer, "Preface," *Haa Kusteeyí, Our Culture*, xxi; LaDuke, *All Our Relations*, 99; and McNabb, "Native Claims in Alaska."

34. Bigjim, "The Poet and Society," 48.

35. Roorda, *Dramas of Solitude*, 5.

36. "Afterword," in Tallmountain, *A Quick Brush of Wings*.

37. Bruchac, "We Are the Inbetweens: An Interview with Mary TallMountain," 13; Costello, "Introduction," in TallMountain, *Listen to the Night*, 8. Further references to this collection will be incorporated parenthetically into the text as *LN*.

38. TallMountain, *The Light on the Tent Wall*, 65. Further references to this collection will be incorporated parenthetically into the text as *LTW*.

39. See Lincoln, *Sing with the Heart of a Bear*, 282.

40. Ibid.

41. For more information on TallMountain's career, contact the TallMountain Circle, P.O. Box 423115, San Francisco, CA 94142. See also Welford, "Mary TallMountain's Writings."

42. Welford, "Reflections on Mary TallMountain's Life and Writings," 62; see also TallMountain, "You *Can* Go Home Again."

43. Lincoln, *Sing with the Heart of a Bear*, 284.

44. TallMountain, *The Light on the Tent Wall*, 48.

45. For more on this reunion, see Lincoln, *Sing with the Heart of a Bear*, 287; and TallMountain, "You *Can* Go Home Again," 9.

46. Bruchac, "We Are the Inbetweens," 15.

47. Ibid., 19.

48. Quoted in LaDuke, *All Our Relations*, 90.

49. Ibid., 1.

50. Rosaldo, *Culture and Truth*, 68–70.

51. For a discussion of Indians as original ecologists, see Martin, "The American Indian as Miscast Ecologist"; and Redford, "The Ecologically Noble Savage." Waller also critiques the ways American Indians are used by white environmentalists, who often reduce them to a generic sameness, appealing to Indians of the past, not the present, which, he argues, ultimately "uncultures them." For more on these points, see Waller, "Friendly Fire."

52. LaDuke, *All Our Relations*, 88.

Conclusion

1. *Northern Exposure*'s Joel Fleischman is not alone in expressing this senti-ment. In the 1990s, Alaska emerged as an object of parody in films such as *In the Line of Fire* (1993), where Clint Eastwood's character is threatened with being sent to this remote outpost as punishment for failing to perform well in his job.

2. For a discussion that contextualizes *Northern Exposure*'s popularity, see Price, *Flight Maps*, 233.

3. For more on the greening of American television, see Price, *Flight Maps*, 234.

4. Producers Falsey and Brand make this comment in the "The First Epi-sode," *Northern Exposure*, CBS (1990).

5. For more on this point, see Fienup-Riordan, *Freeze Frame*, 187–89.

6. For further discussion of *Smoke Signals* as a parody of *Northern Exposure* and the popular Western in general, see Kollin, "Dead Man, Dead West."

7. See "The First Episode," *Northern Exposure*, CBS (1990).

8. Fienup-Riordan, *Freeze Frame*, 189.

9. Romeyn and Kugelmass, "Writing Alaska, Writing the Nation," 263.

10. McPhee, *Coming into the Country*, 149, 152.

11. McGinniss, *Going to Extremes*, 111–12.

12. Muir also wrote an account of his experiences with a dog in Alaska in *Stickeen*.

13. Nelson, *The Island Within*, xiii. Further references will be incorporated parenthetically into the text.

14. Mazel, *American Literary Environmentalism*, 23.

15. Talbot, "The Wilderness Narrative and the Cultural Logic of Capitalism," 330.

16. White, "The Current Weirdness in the West," 13.

17. Guha, *Environmentalism*, 56.

18. Ibid., 69. For a related discussion of Deep Ecology from a non-Western perspective, see Guha, "Deep Ecology Revisited."

19. For another discussion of this point, see Callicott, "Should Wilderness Areas Become Biodiversity Reserves?," 587.

20. Cronon, "Introduction," *Uncommon Ground*, 37.

21. Snyder, "The Rediscovery of Turtle Island," 643.

22. Ibid., 645.

23. Mazel, *American Literary Environmentalism*, xiii.

24. Ibid., xviii.

25. Van Wyck, *Pilgrims in the Wilderness*, 103.

26. Here I draw on the arguments of *Nation* columnist Alexander Cock-burn. For further discussion of Alaska, wilderness politics, and the oil indus-try's latest strategies, see Cockburn, "Dancing with Wolves."

BIBLIOGRAPHY

Abbey, Edward. *Desert Solitaire: A Season in the Wilderness*. New York, 1968.
Reprint. New York: Touchstone, 1990.

Andrews, Byron. "Alaska and Its Gold Fields, and How to Get There." *National Tribune Library* 1, no. 22 (September 18, 1897): 4.

Andrews, Susan B., and John Creed, eds. *Authentic Alaska: Voices of Its Native Writers*. Lincoln: University of Nebraska Press, 1998.

Apter, Emily, and William Pietz, eds. *Fetishism as Cultural Discourse*. Ithaca, N.Y.: Cornell University Press, 1993.

Armbruster, Karla. "Creating the World We Must Save: The Paradox of Television Nature Documentaries." In *Writing the Environment: Ecocriticism and Literature*, edited by Richard Kerridge and Neil Sammells, 218–38. London and New York: Zed Books, 1998.

Armstrong, Nancy. *Desire and Domestic Fiction: A Political History of the Novel*. Oxford and New York: Oxford University Press, 1987.

Atwood, Margaret. *Strange Things: The Malevolent North in Canadian Literature*. Oxford: Clarendon Press, 1995.

Baker, Steve. *Picturing the Beast: Animals, Identity, and Representation*. Manchester and New York: Manchester University Press, 1993.

Beach, Rex. *The Spoilers*. New York: Harper and Brothers, 1905.

——. *The Winds of Chance*. New York: Harper and Brothers, 1918.

——. *The World in His Arms*. New York: Putnam, 1945.

Bederman, Gail. *Manliness and Civilization: A Cultural History of Gender and Race in the United States, 1880–1917*. Chicago: University of Chicago Press, 1995.

Bennett, Jane, and William Chaloupka, eds. *In the Nature of Things: Language, Politics, and the Environment*. Minneapolis: University of Minnesota Press, 1993.

Bercovich, Sacvan. *The American Jeremiad*. Madison: University of Wisconsin Press, 1978.

Berger, John. *About Looking*. New York: Pantheon, 1980.

Berton, Pierre. *Hollywood's Canada: The Americanization of Our National Image*. Toronto: McClelland and Stewart, 1975.

——. *The Klondike Fever: The Life and Death of the Last Great Gold Rush*. New York: Alfred A. Knopf, 1967.

Bigjim, Fred. "The Poet and Society." *Wicazo Sa Review* 5 (1989): 47–48.

Birkland, Thomas. *After Disaster: Agenda Setting, Public Policy, and Focusing Events*. Washington, D.C.: Georgetown University Press, 1997.

Bloom, Lisa. *Gender on Ice: American Ideologies of Polar Expeditions*. Minneapolis: University of Minnesota Press, 1993.

Boime, Albert. *The Magisterial Gaze: Manifest Destiny and American Landscape Painting, c. 1830–1865. Washington, D.C.: Smithsonian Institute Press, 1991.

Bousé, Derek. *Wildlife Films*. Philadelphia: University of Pennsylvania Press, 2000.

Bramwell, Anna. *Ecology in the Twentieth Century: A History*. New Haven, Conn.: Yale University Press, 1989.

Breton, Mary Joy. *Women Pioneers for the Environment*. Boston: Northeastern University Press, 1998.

Broun, Elizabeth. "Foreword," in *The West as America: Reinterpreting Images of the Frontier, 1820–1920*, edited by William Truettner, vii–ix. Washington, D.C., and London: Smithsonian Institution Press, 1991.

Brown, Cathy. "Feds Put Gull-Egg Gathering on Ice." *Juneau Empire*, 7 June 1999, 1.

Bruchac, Joseph W., III. "We Are the Inbetweens: An Interview with Mary TallMountain." *SAIL: Studies in American Indian Literature* 1, no. 1 (1989): 13–21.

——, ed. *Raven Tells Stories: An Anthology of Alaska Native Writing*. Greenfield Center, N.Y.: Greenfield Review Press, 1991.

Buell, Lawrence. *The Environmental Imagination: Thoreau, Nature Writing, and the Formation of American Culture*. Cambridge, Mass.: Belknap/Harvard University Press, 1995.

Bullard, Robert D. "Environmental Blackmail in Minority Communities." In *Reflecting on Nature: Readings in Environmental Philosophy*, edited by Lori Gruen and Dean Jamieson, 132–41. Oxford: Oxford University Press, 1994.

——, ed. *Unequal Protection: Environmental Justice and Communities of Color*. San Francisco: Sierra Club Books, 1994.

Burg, Amos. "The Crislers—Brooks Range Naturalists." *Alaska Fish and Game* 17 (January/February 1985): 9–11.

Burnham, Michelle. *Captivity and Sentiment: Cultural Exchange in American Literature, 1682–1861*. Hanover, N.H.: University Press of New England, 1997.

Burnham, Philip. *Indian Country, God's Country: Native Americans and the National Parks*. Cavello, Calif., and Washington, D.C.: Island Press, 2000.

Burroughs, John, et al., eds. *Alaska: The Harriman Expedition, 1899*. New York, 1901. Reprint. Mineola, N.Y.: Dover, 1986.

Buscombe, Edward, ed. *The BFI Companion to the Western*. New York: Atheneum, 1988.

Buzard, James. *The Beaten Path: European Tourism, Literature, and the Ways to Culture, 1800–1918*. Oxford: Clarendon Press, 1993.

Callicott, J. Baird. "Should Wilderness Areas Become Biodiversity Reserves?." In *The Great New Wilderness Debate*, edited by J. Baird Callicott and Michael P. Nelson, 585–94.

Camacho, David E., ed. *Environmental Injustices, Political Struggles: Race, Class, and the Environment*. Durham, N.C.: Duke University Press, 1998.

Campbell, Robert. "Facing North: Jack London's Imagined Indians on the Klondike Frontier." *Northern Review* 19 (Winter 1998): 122–40.

Carmack, George. *My Experiences in the Klondike*. Published privately by Marguerite Carmack, 1933.

Carter, Debbie. "Film Frames Life of Conservationist." *Fairbanks Daily News-Miner*, 2 November 1997, H8.

Catton, Theodore. *Inhabited Wilderness: Indians, Eskimos, and National Parks in Alaska*. Albuquerque: University of New Mexico Press, 1997.

Cawelti, John G. *The Six-Gun Mystique Sequel*. Bowling Green, Ohio: Bowling Green University Popular Press, 1999.

Chaloupka, William, and R. McGreggor Cawley. "The Great Wild Hope: Nature, Environmentalism, and the Open Secret." In *In the Nature of Things: Language, Politics, and the Environment*, edited by Jane Bennett and William Chaloupka, 3–23. Minneapolis: University of Minnesota Press, 1993.

Chow, Rey. *Writing Diaspora: Tactics of Intervention in Contemporary Cultural Studies*. Bloomington: Indiana University Press, 1993.

Clarke, George Elliot. "White Like Canada." *Transition: An International Review* 7, no. 1 (1997): 98–109.

Clifford, James. "On Collecting Art and Culture." In *Out There: Marginalization and Contemporary Cultures*, edited by Russell Ferguson, Martha Gever, Trinh T. Minh-ha, and Cornel West, 141–69. Cambridge, Mass.: MIT Press, 1990.

Coates, Kenneth S. *Best Left as Indians: Native-White Relations in the Yukon Territory, 1840–1973*. Montreal and Kingston: McGill-Queens University Press, 1991.

——. "The Rediscovery of the North: Towards a Conceptual Framework for the Study of Northern/Remote Regions." *Northern Review* 12/13 (Summer/Winter 1993–94): 15–43.

Coates, Kenneth S., and William R. Morrison. *The Alaska Highway during World War II: The U.S. Army Occupation of Canada's Northwest*. Norman: University of Oklahoma Press, 1992.

Cockburn, Alexander. "Dancing with Wolves." *Nation* 272, no. 8 (26 February 2001): 9.

Cook, James. *The Voyages of Captain James Cook Round the World*. London: Richard Phillips, 1809.

Cook-Lynn, Elizabeth. *Why I Can't Read Wallace Stegner and Other Essays: A Tribal Voice*. Madison: University of Wisconsin Press, 1996.

Cosgrove, Denis. "Habitable Earth: Wilderness, Empire, and Race in America." In *Wild Ideas*, edited by David Rothenberg, 27–41. Minneapolis: University of Minnesota Press, 1995.

Crapol, Edward P., and Howard Schonberger. "The Shift to Global Expansion, 1865–1900." In *From Colony to Empire: Essays in the History of American Foreign Relations*, edited by William Appleman Williams, 135–202. New York: John Wiley and Sons, 1972.

Crisler, Lois. *Arctic Wild*. New York: Harper and Brothers, 1958.

———. *Captive Wild*. New York: Harper and Brothers, 1968.

Cronon, William. "Introduction: In Search of Nature." In *Uncommon Ground: Toward Reinventing Nature*, edited by William Cronon, 23–56. New York: W. W. Norton, 1996.

———. "The Trouble with Wilderness: Or, Getting Back to the Wrong Nature." In *Uncommon Ground: Toward Reinventing Nature*, edited by William Cronon, 69–90.

———. "The Trouble with Wilderness: A Response." *Environmental History* 1, no. 1 (January 1996): 47–55.

———, ed. *Uncommon Ground: Toward Reinventing Nature*. New York: W. W. Norton, 1996.

Cruikshank, Julie, with Angela Sidney, Kitty Smith, and Annie Ned. *Life Lived Like a Story: Life Stories of Three Yukon Native Elders by Julie Cruikshank*. Lincoln: University of Nebraska Press, 1990.

Curwood, James Oliver. *The Alaskan: A Novel of the North*. New York: Grosset and Dunlap, 1922.

———. *Son of the Forests: An Autobiography*. Garden City, N.Y.: Doubleday, Doran and Company, 1930.

Daley, Patrick, with Dan O'Neill. " 'Sad Is Too Mild a Word': Press Coverage of the *Exxon Valdez* Oil Spill." *Journal of Communication* 41, no. 4 (Autumn 1991): 42–57.

Darnovsky, Marcy. "Stories Less Told: Histories of the U.S. Environmentalism." *Socialist Review* 22, no. 4 (October/November 1992): 11–54.

Dauenhauer, Nora Marks. *The Droning Shaman*. Haines, Alaska: Black Current Press, 1988.

———. *Life Woven with Song*. Tucson: University of Arizona Press, 2000.

Dauenhauer, Nora Marks, and Richard Dauenhauer, eds. *Haa Ḵusteeyí, Our Culture: Tlingit Life Stories*. Seattle: University of Washington Press, and Juneau, Alaska: Sealaska Heritage Foundation, 1994.

———, eds. *Haa Shuká, Our Ancestors: Tlingit Oral Narratives*. Seattle: University of Washington Press, and Juneau, Alaska: Sealaska Heritage Foundation, 1987.

———, eds. *Haa Tuwunáagu Yís, for Healing Our Spirit: Tlingit Oratory*. Seattle: University of Washington Press, and Juneau, Alaska: Sealaska Heritage Foundation, 1990.

Davis, Mike. "Dead West: Ecocide in Marlboro Country." *New Left Review* 200 (July/August 1993): 49–73.

———. *Ecology of Fear: Los Angeles and the Imagination of Disaster*. New York: Metropolitan Books, 1998.

Davis, Robert. *Soul Catcher*. Sitka, Alaska: Raven's Bones Press, 1986.

"Decision Criticised." *Dawson Daily News*, 21 October 1903, 1.

DeRoos, Robert. "The Magic Worlds of Walt Disney." In *Disney Discourse: Producing the Magic Kingdom*, edited by Eric Smoodin, 48–68. New York: Routledge, 1994.

Dickey, James. *Deliverance*. New York: Dell, 1970.

Doyle, James. *North of America: Images of Canada in the Literature of the United States, 1775–1900*. Toronto: ECW Press, 1983.

Drinnon, Richard. *Facing West: The Metaphysics of Indian-Hating and Empire-Building*. Minneapolis, 1980. Reprint. New York: Schocken, 1990.

Dryzek, John S. *The Politics of the Earth: Environmental Discourses*. New York: Oxford University Press, 1997.

Duncan, James S., and Nancy G. Duncan. "Ideology and Bliss: Roland Barthes and the Secret History of Landscapes." In *Writing Worlds: Discourse, Text, and Metaphor in the Representation of Landscape*, edited by Trevor J. Barnes and James S. Duncan, 18–37. New York: Routledge, 1992.

Dyer, Richard. *Heavenly Bodies: Film Stars and Society*. New York: St. Martin's Press, 1986.

Egan, Timothy. "Alaskans Don't Want to Be Anyone's Siberia." *New York Times*, 19 December 1993, 3.

———. "Everyone Is Always on Nature's Side: People Just Can't Agree on What's Natural and What's Not." *New York Times*, 19 December 1998, D7.

———. "A Land So Wild, It's Getting Crowded." *New York Times*, 20 August 2000, section 5, p. 10.

Elder, Glen, Jennifer Wolch, and Jody Emel. "*Le Practique Sauvage*: Race, Place, and the Human-Animal Divide." In *Animal Geographies: Place, Politics, and Identity in the Nature-Culture Borderlands*, edited by Jennifer Wolch and Jody Emel, 72–90. London and New York: Verso, 1998.

Ellis, Reuben Joseph. "A Geography of Vertical Margins: Twentieth-Century Mountaineering Essays and the Landscapes of Neoimperialism." Ph.D. diss., University of Colorado, 1991.

Fabian, Johannes. *Time and the Other: How Anthropology Makes Its Object*. New York: Columbia University Press, 1983.

Ferrell, Nancy Warren. *Barrett Willoughby: Alaska's Forgotten Lady*. Fairbanks: University of Alaska Press, 1994.

Fienup-Riordan, Ann. *Freeze Frame: Alaska Eskimos in the Movies*. Seattle: University of Washington Press, 1995.

Forbes, Jack D. "Nature and Culture: Problematic Concepts for Native Americans." *Ayaangwaamizin: The International Journal of Indigenous Philosophy* 1, no. 2 (1997): 3–22.

Foreman, Dave. "Wilderness Areas for Real." In *The Great New Wilderness Debate*, edited by J. Baird Callicott and Michael P. Nelson, 395–407. Athens: University of Georgia Press, 1998.

Foucault, Michel. "Questions on Geography." In *Power/Knowledge: Selected Interviews and Other Writings, 1972–1977*, edited by Colin Gordon, 63–77. New York: Pantheon, 1980.

Frow, John. "Tourism and the Semiotics of Nostalgia." *October* 57 (Summer 1991): 123–51.

Gaard, Greta. "Living Interconnections with Animals and Nature." In *Eco-

feminism: Women, Animals, Nature, edited by Greta Gaard, 1–12. Philadelphia: Temple University Press, 1993.

Gallagher, Carole. *American Ground Zero: The Secret Nuclear War*. New York: Random House, 1993.

George, Rosemary Marangoly. "Homes in the Empire, Empires in the Home." *Cultural Critique* 26 (Winter 1993–94): 95–127.

Georgi-Findlay, Brigitte. *The Frontiers of Women's Writing: Women's Narratives and the Rhetoric of Westward Expansion*. Tucson: University of Arizona Press, 1996.

Glover, James. *A Wilderness Original: The Life of Bob Marshall*. Seattle: The Mountaineers, 1986.

Goetzmann, William H., and Kay Sloan. *Looking Far North: The Harriman Expedition to Alaska, 1899*. Princeton, N.J.: Princeton University Press, 1982.

Gottlieb, Robert. *Forcing the Spring: The Transformation of the American Environmental Movement*. Covelo, Calif., and Washington, D.C.: Island Press, 1993.

Grace, Sherrill. "Comparing Mythologies: Ideas of West and North." In *Borderlands: Essays in Canadian-American Relations*, edited by Robert Lecker, 243–62. Toronto: ECW Press, 1991.

Graebner, Norman, ed. *Manifest Destiny*. Indianapolis: Bobbs-Merrill, 1968.

Green, Martin. *The Great American Adventure*. Boston: Beacon Press, 1984.

Gross, Konrad. "From American Western to Canadian Northern: Images of the Canadian North in Early Twentieth-Century Popular Fiction in Canada." In *Das Natur/Kultur- Paradigma in der englischsprachigen Erzäahlliteratur des 19. Und 20. Jahrhunderts*, edited by Konrad Gross, Kurt Müller, and Meinhard Winkgens, 354–66. Tübingen: Gunter Nerr Verlag, 1994.

Grove, Richard. *Green Imperialism: Colonial Expansion, Tropical Island Edens, and the Origins of Environmentalism, 1600–1860*. Cambridge: Cambridge University Press, 1995.

Guha, Ramachandra. "Deep Ecology Revisited." In *The Great New Wilderness Debate*, edited by J. Baird Callicott and Michael P. Nelson, 271–79. Athens: University of Georgia Press, 1998.

———. *Environmentalism: A Global History*. New York: Longman, 2000.

———. "Radical American Environmentalism and Wilderness Preservation: A Third World Critique." *Environmental Ethics* 11, no. 1 (1989): 71–83.

Hamilton, Cynthia. "Environmental Consequences of Urban Growth and Blight." In *Toxic Struggles: The Theory and Practice of Environmental Justice*, edited by Richard Hofrichter, 67–75. Philadelphia: New Society Publishers, 1993.

Haraway, Donna. *How Like a Leaf: An Interview with Thyrza Nichols Goodeve*. New York: Routledge, 2000.

———. *Primate Visions: Gender, Race, and Nature in the World of Modern Science*. New York: Routledge, 1989.

——. *Simians, Cyborgs, and Women: The Reinvention of Nature.* New York: Routledge, 1991.

Harley, J. B. "Maps, Knowledge, Power." In *The Iconology of Landscape: Essays on the Symbolic Representation, Design, and Use of Past Environments,* edited by Denis Cosgrove and Stephen Daniels, 277–312. Cambridge: Cambridge University Press, 1988.

Harris, A. C. *Alaska and the Klondike Gold Fields.* Washington, D.C.: Library of Congress, 1897.

Harvey, David. *Justice, Nature, and the Geography of Difference.* Cambridge, Mass.: Blackwell, 1996.

Hays, Samuel P. *Conservation and the Gospel of Efficiency.* Cambridge, Mass.: Harvard University Press, 1959.

Hays, Samuel P., with Barbara D. Hays. *Beauty, Health, and Permanence: Environmental Politics in the United States, 1955–1985.* New York: Cambridge University Press, 1987.

Hearing before the Merchant Marine and Fisheries Subcommittee of the Committee on Interstate and Foreign Commerce. Washington, D.C.: U.S. Government Printing Office, 1959.

Hendricks, Helen. "I Teach French in Alaska." *French Review* 16, no. 2 (December 1942): 144–48.

Hepler, John. "Michigan's Forgotten Son: James Oliver Curwood." *Midwestern Miscellany* 7 (1979): 25–33.

Heyne, Eric. "The Lasting Frontier: Reinventing America." In *Desert, Garden, Margin, Range: Literature on the American Frontier,* edited by Eric Heyne, 3–15. New York: Twayne Publishers, 1992.

Hochman, Jhan. *Green Cultural Studies: Nature in Film, Novel, and Theory.* Moscow: University of Idaho Press, 1998.

Hofrichter, Richard. "Introduction." In *Toxic Struggles: The Theory and Practice of Environmental Justice,* edited by Richard Hofrichter, 1 10. Philadelphia: New Society Publishers, 1993.

——, ed. *Toxic Struggles: The Theory and Practice of Environmental Justice.* Philadelphia: New Society Publishers, 1993.

Honderich, John. *Arctic Imperative: Is Canada Losing the North?* Toronto: University of Toronto Press, 1987.

Hornung, Alfred. "Evolution and Expansion in Jack London's Personal Accounts: *The Road* and *John Barleycorn.*" In *An American Empire: Expansionist Cultures and Policies, 1881–1917,* edited by Serge Ricard, 197–213. Aix-en-Provence: Universite de Provence, 1990.

Hughes, Agnes Lockhart. "Honeymooning in Alaskan Wilds." *Overland* 63 (May 1914): 481–85.

Hutchinson, W. H. "The Cowboy in Short Fiction." In *A Literary History of the American West,* edited by Thomas J. Lyon et al., 515–22. Fort Worth, Tex.: Texas Christian University Press, 1987.

Jackson, Kathy Merlock. *Walt Disney: A Bio-Bibliography.* Westport, Conn.: Greenwood Press, 1993.

Jakle, John. *The Tourist: Travel in Twentieth-Century North America*. Lincoln: University of Nebraska Press, 1985.

James, Susie. "Glacier Bay History." In *Haa Shuká, Our Ancestors: Tlingit Oral Narratives*, edited by Nora Marks Dauenhauer and Richard Dauenhauer, 245–59. Seattle: University of Washington Press, and Juneau, Alaska: Sealaska Heritage Foundation, 1987.

Jettmar, Karen. "A Meeting of Generations." *Fairbanks Daily News-Miner*, 2 November 1997, H11.

Jones, Tamara. "Alaska's Trouble on Oily Waters." *Travel and Leisure* (July 1989): 48–54.

Kaplan, Amy. " 'Left Alone with America': The Absence of Empire in the Study of American Culture." In *Cultures of United States Imperialism*, edited by Amy Kaplan and Donald Pease, 3–21. Durham, N.C.: Duke University Press, 1993.

———. "Manifest Domesticity." *American Literature* 70, no. 3 (September 1998): 581–606.

———. "Romancing the Empire: The Embodiment of American Masculinity in the Popular Historical Novel of the 1890s." *American Literary History* 2, no. 4 (Winter 1990): 659–89.

Keen, Dora. "Woman of the Wilderness." *Harper's Weekly* 58 (April 11, 1914): 24–26.

Keller, Robert H., and Michael F. Turek. *American Indians and National Parks*. Tucson: University of Arizona Press, 1998.

Kerridge, Richard. "Small Rooms and the Ecosystem: Environmentalism and DeLillo's *White Noise*." In *Writing the Environment: Ecocriticism and Literature*, edited by Richard Kerridge and Neil Sammells, 182–95. London and New York: Zed Books, 1998.

Kiernan, V. G. *America: The New Imperialism, from White Settlement to World Hegemony*. London: Zed Press, 1978.

King, Margaret J. "The Audience in the Wilderness: The Disney Nature Films." *Journal of Popular Film and Television* 24, no. 2 (Summer 1996): 60–68.

Knobloch, Frieda. *The Culture of Wilderness: Agriculture as Colonization in the American West*. Chapel Hill: University of North Carolina Press, 1996.

Kollin, Susan. "Dead Man, Dead West." *Arizona Quarterly* 56, no. 3 (Autumn 2000): 125–54.

———. " 'The First White Women in the Last Frontier': Writing Nature, Race, and Empire in Alaska Travel Narratives." *Frontiers: A Journal of Women Studies*. Special issue on "Intersections between Feminisms and Environmentalisms," edited by Noël Sturgeon. 18, no. 2 (1997): 105–24.

Kolodny, Annette. "Turning the Lens on 'The Panther Captivity': A Feminist Exercise in Practical Criticism." *Critical Inquiry* 8, no. 2 (Winter 1981): 329–45.

Krakauer, Jon. *Into the Wild*. New York: Villard, 1996.

Kuehls, Thom. *Beyond Sovereign Territory: The Space of Ecopolitics*. Minneapolis: University of Minnesota Press, 1996.

Kupfer, Charles. "The Cold War West as Symbol and Myth: Perspectives from Popular Culture." In *The Cold War American West, 1945–1989*, edited by Kevin J. Fernlund, 167–88. Albuquerque: University of New Mexico Press, 1998.

Kusz, Natalie. *Road Song: A Memoir*. New York: Harper Perennial, 1991.

Labor, Earle. "Jack London." In *A Literary History of the American West*, edited by Thomas J. Lyon et al., 381–97. Fort Worth, Tex.: Texas Christian University Press, 1987.

LaDuke, Winona. *All Our Relations: Native Struggles for Land and Life*. Boston: South End Press, 1999.

LaFeber, Walter. *The New Empire: An Interpretation of American Expansion, 1860–1898*. Ithaca, N.Y.: Cornell University Press, 1963.

Lee, Molly. "Zest or Zeal?: Sheldon Jackson and the Commodification of Alaska Native Art." In *Collecting Native America, 1870–1960*, edited by Shepard Krech III and Barbara A. Hail, 25–42. Washington, D.C.: Smithsonian Institution Press, 1999.

Leed, Eric J. *The Mind of the Traveler: From Gilgamesh to Global Tourism*. New York: Basic Books, 1991.

Leithead, J. Edward. "Tales of Klondike Gold in Dime Novels." *Dime Novel Round-Up* 25, no. 10 (15 October 1957): 88–92.

Leopold, Aldo. *A Sand County Almanac and Sketches Here and There*. New York, 1949. Reprint. New York: Oxford, 1989.

Le Sueur, Meridel. *The Girl*. Albuquerque, N.Mex.: West End Press, 1978.

Lethcoe, Jim, and Nancy Lethcoe. *History of Prince William Sound, Alaska*. Valdez, Alaska: Prince William Sound Books, 1994.

Lewis, R. W. B. *The American Adam: Innocence, Tragedy, and Tradition in the Nineteenth Century*. Chicago: University of Chicago Press, 1955.

Limerick, Patricia Nelson. *Something in the Soil: Legacies and Reckonings in the New West*. New York: W. W. Norton, 2000.

Lincoln, Kenneth. *Sing with the Heart of a Bear: Fusions of Native and American Poetry, 1890–1999*. Berkeley: University of California Press, 2000.

London, Jack. *Burning Daylight*. New York, 1910. Reprint. New York: Macmillan, 1913.

——. *The Call of the Wild*. New York, 1903. Reprint. New York: Penguin, 1960.

——. *The Complete Short Stories of Jack London*. 3 vols. Edited by Earle Labor, Robert C. Leitz III, and I. Milo Shepherd. Stanford, Calif.: Stanford University Press, 1993.

——. *Daughter of the Snows*. New York: Grosset and Dunlap, 1902.

——. *Revolution and Other Essays*. New York: Macmillan Company, 1912.

——. *Smoke Bellew*. New York, 1912. Reprint. New York: Dover Press, 1992.

Lopez, Barry. *Arctic Dreams: Imagination and Desire in a Northern Landscape*. New York: Bantam, 1986.

Lord, Nancy. *Survival*. Minneapolis: Coffee House Press, 1991.

Luke, Timothy W. "Green Consumerism: Ecology and the Ruse of Recycling." In *In the Nature of Things: Language, Politics, and the Environment*, edited by Jane Bennett and William Chaloupka, 154–72. Minneapolis: University of Minnesota Press, 1993.

Lyon, Thomas J., et al., eds. *A Literary History of the American West*. Fort Worth: Texas Christian University Press, 1987.

McCarthy, Max. "Zhirinovsky Upsets D.C." *Buffalo News*, 26 December 1993, 5.

McClintock, Anne. *Imperial Leather: Race, Gender, and Sexuality in the Colonial Contest*. New York: Routledge, 1995.

McClure, John. *Late Imperial Romance*. London: Verso, 1994.

McGinniss, Joe. *Going to Extremes*. New York: Plume, 1980.

McMurtry, Larry. *The Last Picture Show*. New York: Dell, 1966.

McNabb, Steven. "Native Claims in Alaska: A Twenty-Year Review." *Études/Inuit/Studies* 16, nos. 1–2 (1992): 85–95.

McPhee, John. *Coming into the Country*. New York: Noonday Press, 1977.

Mailer, Norman. *Why Are We in Vietnam?* New York: Putnam, 1967.

Maltby, Richard. "John Ford and the Indians: Or, Tom Doniphon's History Lesson." In *Representing Others: White Views of Indigenous Peoples*, edited by Mick Gidley, 120–44. Exeter: Exeter University Press, 1992.

Marshall, Robert. *Alaska Wilderness: Exploring the Central Brooks Range*. 2d ed. Foreword by A. Starker Leopold. Introduction by George Marshall. Berkeley: University of California Press, 1970.

———. *Arctic Village: A 1930s Portrait of Wiseman, Alaska*. New York, 1933. Reprint. Anchorage, Alaska: University of Alaska Press, 1991.

Martin, Calvin. "The American Indian as Miscast Ecologist." *History Teacher* (February 1981): 243–52.

Martin, Glen. "Alaska's Dumb Dance with Wolves." *San Francisco Chronicle*, 30 November 1992, C12.

Marvin, Amy. "Glacier Bay History." In *Haa Shuká, Our Ancestors: Tlingit Oral Narratives*, edited by Nora Marks Dauenhauer and Richard Dauenhauer, 260–91. Seattle: University of Washington Press, and Juneau, Alaska: Sealaska Heritage Foundation, 1987.

May, Elaine Tyler. *Homeward Bound: American Families in the Cold War Era*. New York: Basic Books, 1988.

Mayer, Melanie. *Klondike Women: True Tales of the 1897–98 Gold Rush*. Athens, Ohio: Swallow Press/Ohio University Press, 1989.

Mazel, Davis. *American Literary Environmentalism*. Athens: University of Georgia Press, 2000.

Merchant, Carolyn. *Earthcare: Women and the Environment*. New York: Routledge, 1996.

Miller, Arthur. *Death of a Salesman*. New York, 1949. Reprint. New York: Penguin, 1998.

Miller, Perry. *Nature's Nation*. Cambridge, Mass.: Belknap Press/Harvard University Press, 1967.

Mitchell, Donald Craig. *Sold American: The Story of Alaska Natives and Their Lands, 1867–1959*. Hanover, N.H.: University Press of New England, 1997.

Mitchell, Lee Clark. *Witnesses to a Vanishing America: The Nineteenth-Century Response*. Princeton, N.J.: Princeton University Press, 1981.

Mitchell, W. J. T. "Imperial Landscapes." In *Landscape and Power*, edited by W. J. T. Mitchell, 5–34. Chicago: University of Chicago Press, 1994.

Mitman, Gregg. *Reel Nature: America's Romance with Wildlife on Film*. Cambridge, Mass: Harvard University Press, 1999.

Mourning Dove. *Cogewea, the Half-Blood: A Depiction of the Great Montana Cattle Range*. Boston, 1927. Reprint. Lincoln: University of Nebraska Press, 1981.

Muir, John. *The Cruise of the Corwin*. Boston, 1917. Reprint. San Francisco: Sierra Club Books, 1993.

——. *Edward Henry Harriman*. 1911. Reprint. Coastal Parks Association, 1978.

——. *Letters from Alaska*. Edited by Robert Engeborg and Bruce Merrill. Madison: University of Wisconsin Press, 1993.

——. *Our National Parks*. San Francisco: Sierra Club Books, 1991.

——. *Stickeen*. New York, 1909. Reprint. Berkeley, Calif.: Heyday Books, 1990.

——. *Travels in Alaska*. Foreword by John Haines. Boston, 1915. Reprint. San Francisco: Sierra Club Books, 1988.

Murie, Margaret. *Two in the Far North*. New York, 1962. Reprint. Edmonds, Wash., and Anchorage Alaska: Alaska Northwest Books, 1978.

Murphy, Patrick. " 'The Whole Wide World Was Scrubbed Clean': The Androcentric Animation of Denatured Disney." In *From Mouse to Mermaid: The Politics of Film, Gender and Culture*, edited by Elizabeth Bell, Lynda Haas, and Laura Sells, 125–36. Bloomington: Indiana University Press, 1995.

Murray, John, ed. *A Republic of Rivers: Three Centuries of Nature Writing from Alaska and the Yukon*. New York: Oxford University Press, 1990.

Nabokov, Vladimir. *Lolita*. New York: Putnam, 1955.

Nash, Roderick. "Tourism, Parks, and the Wilderness Idea in the History of Alaska." *Alaska in Perspective* 4, no. 1 (1981): 1–27.

——. *Wilderness and the American Mind*. Rev. ed. New Haven, Conn.: Yale University Press, 1973.

Naske, Claus-M., and Herman E. Slotnick. *Alaska: A History of the 49th State*. 2d ed. Norman: University of Oklahoma Press, 1987.

Nelson, Klondy, as told to Corey Ford. "I Was a Bride of the Arctic." *Saturday Evening Post* 229 (22 December 1956): 11.

Nelson, Richard. *The Island Within*. New York: Vintage, 1989.

Nerlich, Michael. *The Ideology of Adventure: Studies in Modern Consciousness, 1100–1750*. 2 vols. Minneapolis: University of Minnesota Press, 1987.

Niatum, Duane, ed. *Harper's Anthology of Twentieth-Century Native American Poetry*. New York: Harper Collins, 1988.

Nickerson, Sheila. *Disappearance: A Map*. New York: Harcourt Brace and Company, 1996.

Nielson, Jonathan M. *Armed Forces on a Northern Frontier: The Military in Alaska's History, 1867–1987*. Westport, Conn.: Greenwood Press, 1988.

Noble, David. *The End of American History: Democracy, Capitalism, and the Metaphor of Two Worlds in Anglo-American Historical Writing, 1880–1980*. Minneapolis: University of Minnesota Press, 1985.

Norris, Frank. "The Frontier Gone at Last." *The World's Work* 3, no. 4 (February 1902): 1728–31.

Norwood, Vera. *Made from This Earth: American Women and Nature*. Chapel Hill: University of North Carolina Press, 1993.

O'Neill, Dan. *The Firecracker Boys*. New York: St. Martin's, 1994.

Osborne, Brian S. "The Iconography of Nationhood in Canadian Art." In *The Iconography of Landscape: Essays on the Symbolic Representation, Design, and Use of Past Environments*, edited by Denis Cosgrove and Stephen Daniels. 162–78. Cambridge: Cambridge University Press, 1988.

Peterson, Clell T. "Jack London and the American Frontier." Master's thesis, University of Minnesota, 1951.

Phillips, Dana. "Ecocriticism, Literary Theory, and the Truth of Ecology." *New Literary History* 30, no. 3 (Summer 1999): 577–602.

Plesur, Milton. *America's Outward Thrust: Approaches to Foreign Affairs, 1865–1890*. DeKalb: Northern Illinois University Press, 1971.

Porschild, Charlene. "The Environmental Impact of the Klondike Stampede." Paper presented at the Association for Canadian Studies in the United States Biennial Meeting. Seattle, Wash., 15 November 1995.

———. *Gamblers and Dreamers: Women, Men, and Community in the Klondike*. Vancouver: University of British Columbia Press, 1998.

Price, Jennifer. *Flight Maps: Adventures with Nature in Modern America*. New York: Basic Books, 1999.

Przybylowicz, Donna. "Toward a Feminist Cultural Criticism: Hegemony and Modes of Social Division." *Cultural Critique* 14 (Winter 1989–90): 259–301.

Redford, Kent. "The Ecologically Noble Savage." *Cultural Survival Quarterly* 15, no. 1 (1991): 46–48.

Robinson, Forrest G. *Having It Both Ways: Self-Subversion in Western Popular Classics*. Albuquerque: University of New Mexico Press, 1993.

Rodriguez, Sylvia. "Art, Tourism, and Race Relations in Taos: Toward a Sociology of the Art Colony." In *Discovered Country: Tourism and Survival in the American West*, edited by Scott Norris, 143–60. Albuquerque, N.Mex.: Stone Ladder Press, 1994.

Romeyn, Esther, and Jack Kugelmass. "Writing Alaska, Writing the Nation: *Northern Exposure* and the Quest for a New America." In *"Writing" Nation and "Writing" Region in America*, edited by Theo D'haen and Hans Bertens, 253–67. Amsterdam: Vu University Press, 1996.

Roorda, Randall. *Dramas of Solitude: Narratives of Retreat in American Nature Writing*. Albany, N.Y.: SUNY Press, 1998.

Root, Deborah. *Cannibal Culture: Art, Appropriation, and the Commodification of Difference*. Boulder, Colo.: Westview Press, 1996.

Rosaldo, Renato. *Culture and Truth: The Remaking of Social Analysis*. Boston: Beacon Press, 1989.

Ross, Andrew. *The Chicago Gangster Theory of Life: Nature's Debt to Society*. London: Verso, 1994.

Ross, Ken. *Environmental Conflict in Alaska*. Boulder: University Press of Colorado, 2000.

Ruppert, James. " 'Listen for Sounds': An Introduction to Alaska Native Poets Nora Marks Dauenhauer, Fred Bigjim, and Robert Davis." *Northern Review* 10 (Summer 1993): 86–90.

Said, Edward. *Culture and Imperialism*. New York: Vintage, 1993.

Schama, Simon. *Landscape and Memory*. New York: Vintage, 1996.

Schmitt, Peter J. *Back to Nature: The Arcadian Myth in Urban America*. Foreword by John Stilgoe. New York, 1969. Reprint. Baltimore, Md.: Johns Hopkins University Press, 1990.

Schickel, Richard. *The Disney Version: The Life, Times, Art, and Commerce of Walt Disney*. New York: Simon and Schuster, 1968.

Schroeder, Richard, and Roderick P. Neumann. "Manifest Ecological Destinies: Local Rights and Global Environmental Agendas." *Antipode* 27, no. 4 (1995): 321–24.

Scidmore, Eliza. *Alaska: Its Southern Coast and the Sitkan Archipelago*. Boston: D. Lothrop and Company, 1885.

Seager, Joni. "Creating a Culture of Destruction: Gender, Militarism, and the Environment." In *Toxic Struggles: The Theory and Practice of Environmental Justice*, edited by Richard Hofrichter, 58–66. Philadelphia: New Society Publishers, 1993.

——. *Earth Follies: Coming to Feminist Terms with the Global Environmental Crisis*. New York: Routledge, 1993.

Sherwood, Morgan. *Exploration of Alaska, 1865–1900*. New Haven, Conn., 1965. Reprint. Fairbanks: University of Alaska Press, 1992.

Shiva, Vandana. "The Greening of the Global Reach." In *Global Ecology: A New Arena of Political Conflict,* edited by Wolfgang Sachs, 149–56. London: Zed Books, 1993.

Silko, Leslie Marmon. *Yellow Woman and a Beauty of the Spirit: Essays on Native American Life Today*. New York: Simon and Schuster, 1997.

Slotkin, Richard. *The Fatal Environment: The Myth of the Frontier in the Age of Industrialization, 1800–1890*. Reprint. New York: HarperPerennial, 1994.

——. *Gunfighter Nation: The Myth of the Frontier in Twentieth-Century America*. New York: HarperPerennial, 1993.

Slovic, Scott. "Occupancy and Authenticity in Western Environmental Literature." Paper presented at the American Literature Association Annual Meeting, San Diego, Calif., 30 May 1998.

Smith, Conrad. *Media and Apocalypse: News Coverage of the Yellowstone For-*

est Fires, Exxon Valdez Oil Spill, and Loma Prieta Earthquake. Westport, Conn.: Greenwood Press, 1992.

Snyder, Gary. "The Rediscovery of Turtle Island." In *The Great New Wilderness Debate*, edited by J. Baird Callicott and Michael P. Nelson, 642–51. Athens: University of Georgia Press, 1998.

Soper, Kate. *What Is Nature?: Culture, Politics, and the Non-Human*. Cambridge, Mass.: Blackwell, 1995.

Soulé, Michael E., and Gary Lease, eds. *Reinventing Nature?: Responses to Postmodern Deconstruction*. Washington, D.C., and Covelo, Calif.: Island Press, 1995.

Spence, Mark David. *Dispossessing the Wilderness: Indian Removal and the Making of the National Parks*. New York: Oxford University Press, 1999.

Starr, Kevin. *Americans and the California Dream, 1850–1915*. New York: 1973. Reprint. New York: Oxford University Press, 1986.

Staton, Norman. *A National Treasure or a Stolen Heritage: The Administrative History of Glacier Bay National Park and Preserve with a Focus on Subsistence*. Juneau, Alaska: Sealaska Corporation, 1999. ⟨www.sealaska.com⟩.

Stewart, Susan. *On Longing: Narratives of the Miniature, the Gigantic, the Souvenir, the Collection*. Baltimore, Md., 1984. Reprint. Durham, N.C.: Duke University Press, 1993.

Sturgeon, Noël. *Ecofeminist Natures: Race, Gender, Feminist Theory and Political Action*. New York: Routledge, 1997.

Talbot, Carl. "The Wilderness Narrative and the Cultural Logic of Capitalism." In *The Great New Wilderness Debate*, edited by J. Baird Callicott and Michael P. Nelson, 325–33. Athens: University of Georgia Press, 1998.

TallMountain, Mary. *The Light on the Tent Wall: A Bridging*. Los Angeles: American Indian Studies Center, 1990.

———. *Listen to the Night: Poems for the Animal Spirits of Mother Earth*. Edited by Ben Clarke. Introduction by Kitty Costello. San Francisco: Freedom Voices Press, 1995.

———. *A Quick Brush of Wings*. Introduction by Ben Clarke. San Francisco: Freedom Voices Publications, 1991.

———. "You *Can* Go Home Again: A Sequence." In *I Tell You Now: Autobiographical Essays by Native American Writers*, edited by Brian Swann and Arnold Krupat, 1–13. Lincoln: University of Nebraska Press, 1987.

Tanner, Tony. *The Reign of Wonder: Naivety and Reality in American Literature*. Cambridge: Cambridge University Press, 1965.

Thomas, Margaret. "Stream of Tourists Becomes a Torrent." *Juneau Empire*, 16 August 1994, 9.

Tsing, Anna Lowenhaupt. "Transitions as Translations." In *Transitions, Environments, Translations: Feminisms in International Politics*, edited by Joan W. Scott, Cora Kaplan, and Debra Keates, 253–72. New York: Routledge, 1997.

Turner, Frederick Jackson. *The Frontier in American History*. New York, 1920. Reprint. Tucson: University of Arizona Press, 1982.

Udall, Stewart. *The Quiet Crisis*. New York: Avon, 1964.

Van Alstyne, Richard W. *The Rising American Empire*. Oxford, 1960. Reprint. New York: W. W. Norton, 1974.

Vancouver, George. *A Voyage of Discovery to the North Pacific Ocean and Round the World, 1791–1795*. Edited and with an introduction by W. Kaye Lambe. 4 vols. London, 1798. Reprint. London: The Hakluyt Society, 1984.

Van Wyck, Peter. *Primitives in the Wilderness: Deep Ecology and the Missing Human Subject*. Albany, N.Y.: SUNY Press, 1997.

Walker, Franklin. *Jack London and the Klondike: The Genesis of an American Writer*. 1966. Reprint. San Marino, Calif.: Huntington Library Press, 1994.

Waller, David. "Friendly Fire: When Environmentalists Dehumanize American Indians." *American Indian Culture and Research Journal* 20, no. 2 (1996): 107–26.

Watson, Charles N., Jr. *The Novels of Jack London: A Reappraisal*. Madison: University of Wisconsin Press, 1983.

Webb, Melody. *Yukon: The Last Frontier*. Lincoln: University of Nebraska Press, 1985.

Welford, Gabrielle. "Mary TallMountain's Writings: Healing the Heart— Going Home." *ARIEL: A Review of International English Literature* 25, no. 1 (1994): 136–54.

———. "Reflections on Mary TallMountain's Life and Writing: Facing Mirrors." *SAIL: Studies in American Indian Literatures* 9, no. 2 (Summer 1997): 61– 68.

Westra, Laura, and Peter S. Wenz, eds. *Faces of Environmental Racism: Confronting Issues of Global Justice*. Lanham, Md.: Rowman and Littlefield, 1995.

White, Evelyn C. "Black Women and Wilderness." In *The Stories That Shape Us: Contemporary Women Write about the West*, edited by Teresa Jordan and James R. Hepworth, 376–83. New York: W. W. Norton, 1995.

White, G. Edward. *The Eastern Establishment and the Western Experience: The West of Frederic Remington, Theodore Roosevelt, and Owen Wister*. New Haven, Conn., 1968. Reprint. Austin: University of Texas Press, 1989.

White, Richard. "Are You an Environmentalist or Do You Work for a Living?: Work and Nature." In *Uncommon Ground: Toward Reinventing Nature*, edited by William Cronon, 171–85. New York: W. W. Norton, 1996.

———. "The Current Weirdness in the West." *Western Historical Quarterly* 28, no. 1 (Spring 1997): 5–16.

Whitehead, John. "Alaska and Hawai'i: The Cold War States." In *The Cold War American West, 1945–1989*, edited by Kevin J. Fernlund, 189–210. Albuquerque: University of New Mexico Press, 1998.

———. "Noncontiguous Wests: Alaska and Hawai'i." In *Many Wests: Place, Culture, and Regional Identity*, edited by David M. Wrobel and Michael C. Steiner, 315–41. Lawrence: University Press of Kansas, 1997.

Wilkie, Rab, and the Skookum Jim Friendship Centre. *Skookum Jim: Native and Non-Native Stories and Views about His Life and Times and the Klon-*

dike Gold Rush. Whitehorse, Yukon: Heritage Branch/Department of Tourism, 1992.

Williams, Raymond. *The Country and the City*. New York: Oxford University Press, 1973.

———. *Keywords: A Vocabulary of Culture and Society*. New York: Oxford University Press, 1976.

———. *Problems in Materialism and Culture: Selected Essays*. London: Verso, 1980.

Williams, Terry Tempest. " 'You Have to Know How to Dance': The Inspiration of Margaret Murie." *Anchorage Daily News*, 5 October 1997, F13.

Wilson, Alexander. *The Culture of Nature: North American Landscapes from Disney to the Exxon Valdez*. Cambridge, Mass.: Blackwell, 1992.

Winkler, Karen J. "Inventing a New Field: The Study of Literature about the Environment." *Chronicle of Higher Education*, 9 August 1996, A8.

Winner, Langdon. *The Whale and the Reactor: A Search for Limits in an Age of High Technology*. Chicago: University of Chicago Press, 1986.

Wister, Owen. *The Virginian: A Horseman of the Plains*. 1902. Reprint. New York: Signet, 1979.

Wolch, Jennifer, and Jody Emel. "Preface." In *Animal Geographies: Place, Politics, and Identity in the Nature-Culture Borderlands*, edited by Jennifer Wolch and Jody Emel, xi–xx. London: Verso, 1998.

———, eds. *Animal Geographies: Place, Politics, and Identity in the Nature-Culture Borderlands*. London: Verso, 1998.

Wolfley, Jeanette. "Ecological Risk Assessment and Management: Their Failure to Value Indigenous Traditional Ecological Knowledge and Protect Tribal Homelands." *American Indian Culture and Research Journal* 22, no. 2 (1998): 151–69.

Wooley, Christopher B. "Alutiiq Culture before and after the *Exxon Valdez* Oil Spill." *American Indian Culture and Research Journal* 19, no. 4 (1995): 125–53.

Worster, Donald. *Under Western Skies: Nature and History in the American West*. New York and Oxford: Oxford University Press, 1992.

Wrobel, David M. *The End of American Exceptionalism: Frontier Anxiety from the Old West to the New Deal*. Lawrence: University of Kansas Press, 1993.

Young, Samuel Hall. *Alaska Days with John Muir*. New York, 1915. Reprint. Salt Lake City, Utah: Peregrine Smith Books, 1990.

INDEX

Abbey, Edward, vii

Adirondack Mountains, 41

Alaska: and *Exxon Valdez* oil spill, 2–5, 17–18, 21–22, 50–51, 163, 182 (n. 37); role of in U.S. expansion, 5–12, 14, 21–22, 31–36, 60–71, 72, 89, 104–6, 144–47, 153–55; in U.S. spatial imagination, 5–12, 15–17, 20–22, 24–25, 50–51, 56–57, 61, 92–93, 97–98, 123, 126, 162, 179 (n. 4), 197 (n. 1); and conservation movement, 6–7, 9–11, 45, 70–71, 75–76, 82, 87–88; and tourism, 12–17, 28–29, 30–33, 36–38, 93, 136, 141–42, 150–51, 167; and European exploration, 14–15, 30–31; in nature writing, 24–25, 28–57, 92–94, 98–126, 168–73; and Muir, 24–39; Southeast, 28–32, 39, 131–48; and Young, 30–34; and preservation movement, 33–34; and National Park Service, 39; and Marshall, 39–46; and McPhee, 46–51, 168–69; and Lopez, 51–53; Native environmental sovereignty, 55, 129, 131–60; and Haines, 56–57; and London, 60–74; as entryway into Canada, 61–62, 64–71, 74, 76–81; compared to Africa, 65; compared to India, 65; as Russian America, 75–76; and Beach, 75–81; and Curwood, 81–86; and Soviet threat, 87–89; and L. Crisler, 93–115; and Disney nature films, 94–100, 107–8; and U.S. military, 97–99, 107–9; and Cold War anxieties, 108–9; in *The Thing*, 108–9; and M. Murie, 115–22, 124–26; as inhabited wilderness, 124; and R. Dauenhauer, 131–32; and N. Dauenhauer, 131–40; and Davis, 140–48; and Tall-Mountain, 148–59; in *Northern Exposure*, 162–69; and McGinniss, 169; and Kusz, 170; and Lord, 170; and Nickerson, 170; and Nelson, 170–73; in American literature, 179 (n. 4). *See also* Last Frontier; Wilderness

Alaska Conservation Foundation, 177

Alaska Days with John Muir (Young), 30–34

Alaska: Its Southern Coast and Sitkan Archipelago (Scidmore), 37, 93

Alaskan, The (Curwood), 82–89

Alaska National Interest Lands Conservation Act (ANILCA), 115

Alaska Natives: respond to *Exxon Valdez* oil spill, 3–5; dispossession of, 13, 33, 35, 39, 44–45, 112–15, 123–24, 132, 140, 144–47, 150–58, 194 (n. 84); in contact with European explorers, 14–15; involvement in tourism, 30, 33, 37, 136, 141–42, 150; as guides for Muir, 30, 37; in nature writing, 33–36, 44–45, 111–13, 170; respond to Atomic Energy Commission, 55; and post–World War II population, 99; and TransAlaska Pipeline, 123; and mainstream environmental movement, 123–24, 129, 130–31, 138–40, 153–54, 159, 176–77; and counternarratives of Alaska, 129–60; N. Dauenhauer, 131–40; Davis, 140–48; TallMountain, 148–59; in *Northern Exposure*, 162, 164–69; and McPhee, 168–69; and McGinniss, 169; defined, 187 (n. 89), 194 (n. 5); and wildlife management laws, 194 (n. 81). *See also* Aleuts; American Indians; Athabascans; Eskimos; First Nations; Haida Indians; Tlingits

Alaska Native Claim Settlement Act (ANSCA), 123, 147, 168

Alaska Wilderness (Marshall), 41–45

Aleuts, 4, 129

Aleutian Islands, 98, 148

Alexandrian Archipelago, 30

Allen, Paula Gunn, 149

American Adam, 24, 94, 144, 163–64, 173, 183 (n. 1)

American exceptionalism, 7, 11–12, 19–20, 22, 29, 46, 102, 182 (n. 51)

American Indians: marginalized by mainstream environmental movement, 11, 20, 21, 38, 123–24, 139, 176–77, 185 (n. 41), 196 (n. 51); dispossession of, 38, 63, 68–69, 71–73; and back-to-nature movement, 63; in captivity narratives, 101–2; as original ecologists, 158, 196 (n. 51). *See also* Alaska Natives; Aleuts; Athabascans; Eskimos; First Nations; Haida Indians; Tlingits

Anachronistic space, 92, 112, 113, 114, 159

Anaktuvik Pass, 111

Anchorage, Alaska, 47, 56, 99, 108, 116, 141

Andrews, Byron, 78–79

Animals. *See* Wildlife

Antiquities Act, 32

Anxiety of belatedness, 15, 44, 182 (n. 39)

Arctic, the: role of in national security, 6; and Muir, 34; and Marshall, 39–46; and Lopez, 51–53; and Atomic Energy Commission, 55; and L. Crisler, 94–95, 99–115; and M. Murie, 116–21; and N. Dauenhauer, 136; and oil development, 174, 176–77

Arctic Dreams (Lopez), 46, 51–54

Arctic National Wildlife Refuge (ANWR), 94, 115, 174, 176–77

Arctic Village (Marshall), 41, 45–46

Arctic Wild (L. Crisler), 95, 98, 99, 103, 106–7, 110–13

Armbruster, Karla, 100

Athabascans, 131, 148–59, 170

Atomic Energy Commission, 55

Back-to-nature movement, 63–64, 67, 72, 188 (n. 17)

Beach, Rex: as frontier writer, 60, 62, 75–81, 89, 126, 131; *The Spoilers*, 75; *The World in His Arms*, 75–76; *The Winds of Chance*, 75, 76–78, 79

Berger, John, 110, 113–14

Bering Sea, 88

Bigjim, Fred, 148

Birkland, Thomas, 50–51

Brand, Joshua, 166–67

Brazil, 11, 52

British Columbia, 7, 74

Brooks Range, Alaska, 25, 39, 42, 46, 93, 104, 117, 119

Buell, Lawrence, 29

Bullard, Robert, 146–47

Burnham, Michelle, 102

Burning Daylight (London), 67–73, 81

Bush, George W., 174

Buzard, James, 36–37

California: Muir's experiences in, 33, 34; as portrayed by London, 64, 66, 72; as portrayed by Curwood, 86; and writings of N. Dauenhauer, 136; and writings of TallMountain, 148, 151, 156; and West Coast energy "crisis," 176

Call of the Wild, The (London), 67

Canada: role of in U.S. expansion, 6, 7; and Lopez, 51; as depicted by London, 61–62, 64–74; as depicted by Beach, 76–85; as depicted by Curwood, 81–85; and U.S. literary imagination, 89, 180 (n. 16), 189 (n. 31), 189 (n. 42)

Cape Thompson, Alaska, 55

Captive Wild (L. Crisler), 114

Captivity narratives, 101–2

Carmack, George, 81, 190 (n. 61)

Carmack, Kate (Shaaw Tláa), 81, 190 (n. 61)

Carrighar, Sally, 124

Catton, Theodore, 110, 123, 124

Cawley, R. McGreggor, 23

Chaloupka, William, 23

"Cheechacko," 150

Chilkoot Pass, 70, 85

Chow, Rey, 139–40

Coates, Ken, 80–81, 187 (n. 10)

Cold War, 98, 99, 108–9, 192 (n. 44)

Colonial adventure narratives, 65, 188 (n. 22)

Colonialism: and Alaska's strategic position, 7–8; role of mainstream environmentalism in, 11, 54–55, 129–30; and redemptive unmapping, 14; and frontier ideologies, 19–20, 60–63; and concept of wilderness, 20–21; and Muir, 31–32, 34–35; and London, 61–62, 64–74; and Beach, 75–81; and Curwood, 82, 88–89; and domestic ideologies, 111, 186 (n. 86); in Alaska Native poetry, 144–47, 152–57. *See also* Expansion

Coming into the Country (McPhee), 46–50, 168–69

Conservationism: and frontier ideologies, 6, 9–11; and Muir, 38, 54; and Marshall, 54; and U.S. expansion, 60–61, 70–71, 82, 87–88; and London, 70–71; and Beach, 75–76; and Curwood, 82, 87–88. *See also* Mainstream environmental movement; Preservationism

Continentalism, 74

Cook, Al, 169

Cook, Captain James, 14

Cook-Lynn, Elizabeth, 73

Corwin Expedition, 34–35, 184 (n. 20)

Cosgrove, Denis, 59, 62–63, 73

Country and the City, The (R. Williams), 26

Crisler, Herb, 94, 111–12, 114

Crisler, Lois: and nature writing, 93, 94–95, 98–107, 109–15, 125–26, 131; *Arctic Wild*, 93, 95, 103–4, 106–7, 110–14; *White Wilderness*, 93, 96, 99, 107, 109; *Captive Wild*, 114

Cronon, William, 122–23, 128–29, 161, 175

Cultural studies, 25–28, 177

Curwood, James Oliver: as frontier writer, 60, 62, 81–89, 126, 131; *The Alaskan*, 82–89

Daley, Patrick, 3–5

Darnovsky, Marcy, 38

Dauenhauer, Nora Marks: life and works, 131–40, 159; *Haa Shuká, Our Ancestors*, 131; *Haa Tuwunáagu Yís, for Healing Our Spirit*, 131; *Haa Kusteeyí, Our Culture*, 131; *The Droning Shaman*, 132–36, 138–40; *Life Woven with Song*, 132, 135–38

Dauenhauer, Richard, 131

Davis, Robert: life and works, 131, 140–48; *Soul Catcher*, 140–48

Dawson City, Yukon Territory, Canada, 61, 68, 84, 85, 187 (n. 9)

Dawson Daily News, 80

Deep Ecology, 139

Denali (Mt. McKinley), 13

Denali National Park, Alaska, 24

Disappearance: A Map (Nickerson), 170

Disney, Walt, 94, 97

Disney Company, 112, 133, 134, 191 (n. 14); True-Life Adventure Se-

ries, 93, 94–96, 98–100, 107–9;
Seal Island, 96; *The Living Desert*,
96; *The Vanishing Prairie*, 96;
White Wilderness, 93, 96, 100, 107,
109, 114
Distant Early Warning System
(DEW Line), 98
Domestic ideologies, 103–6, 109, 111–
12, 118–20, 192 (n. 36)
Doyle, James, 89
Droning Shaman, The (N. Dauen-
hauer), 132–36, 138–40

Eagle, Alaska, 49, 77
Earth Day, 18
Ecocriticism, 25–28, 183 (nn. 7, 11),
194 (n. 2)
Ecofeminism, 102–3, 125
Ecology of affluence, 174
"Environmental blackmail," 146–47
Environmental cultural studies, 22,
177–78
Environmental justice movement,
116, 128, 174, 193 (n. 67)
Environmental movement. *See*
Mainstream environmental movement
Environmental Protection Agency, 2
Eskimos: as depicted by Muir, 34–35;
and Marshall, 42, 45; and Lopez,
53; and Atomic Energy Commission, 55; and Disney films, 97, 99;
as depicted by L. Crisler, 111–15;
and M. Murie, 116; and Alaska Native Claims Settlement Act, 123;
and National Park Service, 124;
and countervisions of Alaska, 129;
effects of global tourism on, 136;
and environmentalism, 139; Bigjim, 148; as depicted by McGinniss, 169; and oil development,
176; and Prince William Sound,
179 (n. 1); and term explained, 194
(n. 5)
Expansion: U.S. nation-building and

Alaska, 5, 6, 7, 8, 9, 11, 13, 89; role
of mainstream environmental
movement in, 11, 54–55, 63–64,
129–30; role of nature writing in,
31–33, 34–36; and frontier ideologies, 19–20, 60–63; and concept of wilderness, 20–21, 62–63;
role of Muir in, 31–32, 34–35; role
of London in, 61–62, 64–74; role
of Beach in, 75–81; role of Curwood in, 82, 88–89; and Native
environmental sovereignty, 56,
144–47, 152–57. *See also* Colonialism
Exxon Company, 12–14
Exxon Valdez oil spill: public responses to, 2–5, 17–18, 21–22, 50–
51; tourism affected by, 12–17, 182
(n. 37); and *Northern Exposure*,
163
Eyre, Chris, 166–67

Fairbanks, Alaska, 47, 56, 84, 93, 99,
118, 121
Falsey, John, 166, 167
Feminism, 102–3, 105, 125, 175
Fienup-Riordan, Ann, 167
First Nations, 68–69, 78–81, 190
(n. 62)
Forbes, Jack, 130
Forest Service, U.S., 33, 40, 43–44,
83
Foucault, Michel, 20
Freud, Sigmund, 139–40
Frontier: and American exceptionalism, 19–20; as discursive construction, 19–20, 123–24; as euphemism for empire, 20; threats
to, 28, 69–70; and U.S. expansion,
60, 71, 87, 187 (n. 2); and white
flight, 63; and white masculinity,
63–65; and American national
hero, 68; and London; and Beach,
75–79; and Curwood, 84–85, 86;
and Cold War anxieties, 96–97;

wildlife as symbol of, 110; and
M. Murie, 117. *See also* Last
Frontier

Frontier thesis, 60–61

Frow, John, 36

Gaard, Greta, 102

Geary, Cynthia, 167

Glacier Bay, Alaska, 33, 34, 36, 39, 184
(n. 17), 185 (n. 41)

Glaciers: and Vancouver, 30; and
Muir, 30–33, 35–36, 39; and
Young, 30–33, 184 (n. 17), 185
(n. 41)

Globalization: effects on Alaska, 48–
49, 57, 159, 172; and environmentalism, 52–55, 57, 129–31, 138–39,
172, 182 (n. 57); and Disney icons,
131, 134; tourism affected by, 136

Goetzmann, William, 11

Going to Extremes (McGinniss), 169

Gottlieb, Robert, 40

Green, Martin, 59

Green consumerism, 18–19

Greenland, 53

Grove, Richard, 54

Guha, Ramachandra, 174

Gulf of Alaska, 117

Haines, John, 56

Haggard, H. Rider, 65

Hamilton, Cynthia, 139

Hamilton Bay, Alaska, 144

Haraway, Donna, 13, 27, 112, 113

Harriman, Edward Henry, 9, 35–36

Harriman Alaska Expedition, 35–
37

Harris, A. C., 79

Harvey, David, 22

Hawai'i, 8

Henderson, Robert, 81

Hensley, Willie, 168–69

Heyne, Eric, 20, 24

Hickel, Walter, 109–10

Hochman, Jhan, 26

Hobson, Geary, 149

Hoonah, Alaska, 39

Hudson's Bay Company, 74

Indians. *See* Alaska Natives; Aleuts;
American Indians; Athabascans;
Eskimos; Haida Indians; Tlingits

Inside Passage, 28, 38, 85, 94, 111, 141,
150

Into the Wild (Krakauer), 24

Island Within, The (Nelson), 171–73

Jakle, John, 33

Johnson, Lyndon, 115

Juneau, Alaska, 47, 48, 86, 132, 133–34

Kake, Alaska, 140, 144, 146

Kaplan, Amy, 19–20, 104–5

Kipling, Rudyard, 65, 72

Klondike, the (Yukon Territory), 64,
66, 68, 72, 74, 76, 77, 78–83, 85, 188
(n. 19)

Knobloch, Frieda, 43–44

Koyukon Athabaskans, 148–51, 170

Krakauer, Jon, 24

Kuehls, Thom, 52

Kugelmass, Jack, 167–68

Kusz, Natalie, 170

LaDuke, Winona, 127, 129, 132, 154,
158

LaFeber, Walter, 6–7

Last Frontier: Alaska as, 2, 5, 6, 11, 24,
29, 56, 92, 97, 110, 126, 138; and
tourism industry, 16; and national
ideologies, 19, 21, 22, 57, 62; critiques of, 25, 48, 51, 52, 55, 56, 129,
131, 155, 159, 162, 177; and mainstream environmentalism, 41, 45,
47; and Curwood, 84–85, 86; and
white settlers, 117, 118, 121; threatened by oil development, 123, 168,
169, 174; and *Northern Exposure*,
162, 167. *See also* Alaska; Frontier

Leopold, A. Starker, 44

Letters from Alaska (Muir), 47
Lewis and Clark, 41–42
Life Woven with Song (N. Daunhauer), 132, 135–38
Light on the Tent Wall, The (TallMountain), 150–53
Limerick, Patricia, 71
Lincoln, Kenneth, 150
Listen to the Night (TallMountain), 153–58
Living Desert, The (Disney), 96
Lodge, Henry Cabot, 80
Logging industry, 2, 146–47, 171–72
London, Jack: as frontier writer, 60, 61–73, 88–89, 94, 110, 114, 126, 131, 170, 188 (n. 19); as "Kipling of the Klondike," 64; "God of His Fathers," 64–65; *Smoke Bellew*, 66–67; *The Call of the Wild*, 67; "To the Man on Trail," 67; *Burning Daylight*, 67–73, 81; *Daughter of the Snows*, 70; and alcohol, 189 (n. 34)
Lopez, Barry, 25, 46
Lord, Nancy, 170
Luke, Timothy W., 18

McCandless, Chris, 24
McClintock, Anne, 92, 104, 125
McClure, John, 14
McGinniss, Joe, 169
McPhee, John, 25, 46–50, 103, 131, 168–69
Mainstream environmental movement: and dispossession of American Indians, 11–12, 20, 37, 116, 123, 128–29, 131, 140, 153–54, 159, 176–77; and constructions of wilderness, 20, 116, 194 (n. 2); and Muir, 24, 37; and Marshall, 24, 40; and environmental justice movement, 193 (n. 67). *See also* Conservationism; Preservationism
Marshall, Robert, 24–25, 39–46, 48, 50, 93, 117, 126, 131

Masculinity, 63–64, 68, 71, 92–93, 146, 187 (n. 13)
May, Elaine Tyler, 109
Mazel, David, 161, 173, 175–76
Merchant, Carolyn, 103, 125
Miles, Elaine, 167
Military, United States, 74, 97–99, 107–9, 121, 171
Miller, Perry, 22
Milotte, Al, 97
Mitchell, W. J. T., 1
Mitman, Gregg, 96
Monroe, Marilyn, 12–13, 181 (n. 30)
Morrow, Rob, 167
Muir, John: as nature writer, 24–25, 28–39, 46, 50, 56, 83, 94, 110, 111, 114, 126, 131, 141–42, 150, 183 (n. 17), 184 (n. 20), 185 (n. 34); *Travels to Alaska*, 28–29, 36; *Letters from Alaska*, 30; as depicted by Young, 31–34; *Edward Henry Harriman*, 35–36
Murie, Olaus, 94, 115, 116, 119–20, 124
Murie, Margaret: as nature writer, 93–94, 115–22, 124–26, 131; *Two in the Far North*, 115–16, 118–21, 125–26; *Island Between*, 116; *Wapiti Wilderness*, 116; as "mother" of modern environmental movement, 191 (n. 8)

National Geographic, 96
National parks, 11, 39, 115, 185 (n. 41)
National Park Service, 39, 43, 124
National Petroleum Reserve, 176–77
National security, 6, 87, 88–89, 108–9
Nature: effects of *Exxon Valdez* oil spill on, 2–5; commercial uses of, 5, 10, 12; subsistence uses of, 5, 124; and relation to national identity, 5–6, 11, 21–22, 60, 62–64, 98, 107–9; and U.S. expansion, 5–6, 65, 71, 73; as discursive construction, 13, 17, 19, 21–22, 26–28, 122–23; as

sanctuary, 17, 39, 57, 122–23, 128–30; as public relations strategy for corporations, 18–19, 50–51; treated as religion, 22, 128–29; and cultural studies, 26–28; imagined outside the social realm, 26, 122–23, 130; glaciers, 30–33, 35–36, 39; and trophy hunting, 32, 111, 184 (n. 19); as scenic resource, 34; and frontier thesis, 60–61; and white flight, 63; and white masculinity, 63–65; and white women, 93–94, 118–21, 124–25; post–World War II attitudes toward, 95–96, 116–25; and Disney films, 96–100; and Cold War anxieties, 98, 107–9; as feminized space, 102–3; and domestic ideology, 103–7, 118–21, 124–25; and environmental justice movement, 128–30; American Indian understandings of, 130, 131–60, 170; and logging companies, 146; and *Northern Exposure*, 162–65, 167–69; and Nelson, 170–73. *See also* Frontier; Wilderness; Wildlife

Nature's Nation (Miller), 22

Nature writing: and McPhee, 25, 46–51; and Lopez, 25, 51–54; critical responses to, 25–28, 29; as asocial practice, 27–28, 56; and Muir, 28–39, 183 (n. 5); and Marshall, 39–45, 183 (n. 5); and Alaska Native environmentalism, 56, 148; and Haines, 56; and L. Crisler, 93–115, 125–26; and white women, 93–94, 125–26; and M. Murie, 115–26; and Nelson, 170–73

Nelson, Richard, 171–73

Nerlich, Michael, 43

Neumann, Roderick P., 130

New York, 162, 164, 166

New York Times, 177

Nickerson, Sheila, 170

Noble, David W., 19

Nome, Alaska, 76, 84, 136

Norris, Frank, 60

North Pole, 53, 108–9

North Slope, 2, 116, 123

Northwest Passage, 53, 179 (n. 1)

Northern Exposure, 162–68, 169

Northern Pacific Railroad, 28

Nuclear weapons, 55, 108, 157–58

Nulato, Alaska, 148, 149

Nyby, Christian, 108–9

Oil industry, 2, 176–77, 197 (n. 26)

Oil Pollution Act of 1990, 51

O'Neill, Dan, 55

Orientalist melancholia, 138–39

Outside Magazine, 24

Pacific Ocean, 7, 8

Pease, Donald, 26

Persian Gulf War, 11–12

Pinchot, Gifford, 33, 83–84

Postmodernism, 26–27

Poststructuralism, 27

Preservationism, 32–34, 38, 88, 115–25, 128, 140, 188 (n. 16). *See also* conservationism, mainstream environmental movement

Pribilof Islands, 96

Price, Jennifer, 130

Prince William Sound: in news coverage of *Exxon Valdez* oil spill, 2–5; economic development of, 2–5, 179 (n. 1); and tourist industry, 13; exploration in, 14–15, 35, 179 (n. 1)

Project Chariot, 55

Prudhoe Bay, Alaska, 2, 46, 123, 168, 176

Pryzybylowicz, Donna, 102–3

Raven, 135, 144

Redemptive unmapping, 14

Road Song (Kusz), 170

Robinson, Forrest G., 77

Rodriguez, Sylvia, 34

Romeyn, Esther, 167–68

Roorda, Randall, 27–28, 148
Roos, Robert De, 96
Roosevelt, Theodore, 63–64, 80, 88, 188 (n. 17)
Root, Elihu, 80
Rose, Wendy, 149
Ross, Andrew, 19, 181 (n. 26)
Rowlandson, Mary, 101
Ruppert, James, 145
Russia, 6–7, 9, 21–22, 29, 74, 86–87, 88, 89, 98, 108–9
Russian America, 75–76

St. Augustine, Florida, 86
San Francisco, 66, 72, 75, 86, 149, 156
San Francisco Examiner, 153
Saturday Evening Post, 83
Schama, Simon, 23
Schickel, Richard, 97
Schroeder, Richard, 130
Scidmore, Eliza, 37, 93
Seager, Joni, 54, 130
Seal Island, 96
Seward, William Henry, 7, 74
Sheldon Jackson School, 146
Sherwood, Morgan, 7–9
Shiva, Vandana, 138–39
Sierra Club, 24
Silko, Leslie Marmon, 57
Sitka, Alaska, 75, 121, 146
Skagway, Alaska, 80, 85–86
Skookum Jim, 81, 190 (n. 61)
Sloan, Kay, 11
Slotkin, Richard, 5
Smoke Bellew (London)
Smoke Signals (Dir. Eyre), 166–67
Snyder, Gary, 175
Sonoma Valley, 73
Soper, Kate, 106–7
Soul Catcher (Davis), 140–48
South Franklin Avenue, 133
Soviet Union, 87, 98, 108–9. *See also,* Russia
Spoilers, The (Beach), 76

Starr, Kevin, 66
Sublime, the, 5, 106–7
Superior National Forest, 82
Survival (Lord), 170

Talbot, Carl, 173–74
TallMountain, Mary: life and works of, 127, 131, 148–59; *The Light on the Tent Wall*, 150–53; *Listen to the Night*, 153–58
TallMountain Circle, The, 149, 196 (n. 41)
Tatilek, Alaska, 4
Tenderloin Reflection and Education Center, The, 149
Thing, The (Dir. Nyby), 108–9
Thoreau, Henry David, 46, 47, 186 (n. 61)
Timber industry. *See* Logging industry
Tlàa, Shaaw (Kate Carmack), 81, 190 (n. 61)
Tlingits, 30, 33, 36, 37, 39, 131, 134–36, 138, 140–47
Tourism: Alaska as shaped by, 2, 15–17, 28, 36–37, 133, 136, 150, 167, 181 (n. 37); and crisis management, 12–14; Alaska Native involvement in, 30, 33, 37, 133, 136, 141–42, 150; as depicted by Muir, 36–37; and national parks, 39, 43; as depicted by Marshall, 41–42, 45–46; as depicted by Curwood, 85
TransAlaska Pipeline, 46, 51, 116–17, 123
True-Life Adventure Series. *See* Disney Company, True-Life Adventure Series
Tsing, Anna Lowenhaupt, 54
Tundra Times, 4–5, 55
Turner, Frederick Jackson, 5; and frontier thesis, 60–61
Turner, George, 80
Two in the Far North (M. Murie), 115, 118–21

United Church of Christ Commission for Racial Justice, 128
University of Alaska, Fairbanks, 115, 149
University of Alaska Press, 116
Urban ecology, 47, 149, 156–58
Urban spaces, 17, 22, 47, 56, 149
Urban wilderness, 72, 189 (n. 39)
U.S./Canadian Boundary Dispute, 78–80
United States Military. *See* Military, United States

Valdez, Alaska, 2, 4, 18, 117
Vancouver, George, 15, 30
Vanishing Prairie, The (Disney), 96
van Wyck, Peter, 122, 139, 176

Washington State, 80, 163
Webb, Melody, 11
Western Union Telegraph Company, 9
Westerns, 60, 62, 73, 76, 77, 164–68, 188 (n. 22), 189 (n. 42)
White, Richard, 37, 122, 174
Whitehorse, Yukon Territory, 85
White Wilderness (Disney), 93, 96, 100, 107, 109
Whitman, Walt, 143
Wilderness: Alaska imagined as, 2–3, 5, 11, 17, 21, 24, 41–46, 95, 162; threats to, 2–3, 5, 17, 28; and preservation movement, 10, 40; and conservation movement, 11; role in U.S. expansion, 11, 62–64, 67, 71, 73; as commodity, 12, 17; and tourism industry, 17; as imagined geography, 17, 20, 122–24, 128–29, 173–77; as antisocial space, 20–21, 38, 40, 46; as sanctuary, 46, 122, 128–29, 173–74; as hermetically sealed space, 52, 56, 128–29; as ideology, 53, 122–24, 173–77; and white flight, 63; and white masculinity, 63–65; urban, 72; and

L. Crisler, 94–115; and white women, 92–94, 101–7, 109–22, 124–26; and Cold War anxieties, 98; and domestic ideology, 103–7, 118–21, 124–25; and M. Murie, 115–26; amenity uses of, 116, 118, 122, 125; and Alaska Native environmental sovereignty, 128–60; treated as religion, 128–29; and Nelson, 170–73; as leisure-time concept, 173–74; and environmental fundamentalism, 174; and ecology of affluence, 174; Third World attitudes toward, 174; depicted as Eden, 175; and American Indians, 185 (n. 41); and green romanticism, 194 (n. 2); and oil companies, 197 (n. 26). *See also* Alaska; Last Frontier; Nature
Wilderness Act, the, 115
Wilderness Society, the, 24, 40, 115
Wildlife: effects of *Exxon Valdez* oil spill on, 2, 3, 5, 13; and tourism, 14; sea otter, 15; and expansion, 15, 75; seal, 15, 75, 140; bear, 26–27, 53; wolf, 93, 95, 101, 102, 106, 109–12, 114–15, 157–58; and Crislers, 94, 98–114; and Disney nature films, 94, 98–114; and captivity, 95, 99–102, 106, 109, 111, 114, 191 (n. 12); effects of U.S. military on, 98–99; caribou, 99, 116; lemming, 100; expelled from modern culture, 110; as symbol of the Last Frontier, 110–11; salmon, 134–35, 137–38; whale, 138, 153–54; striped quagga, 155; Bengal tiger, 155; management laws and Alaska Natives, 194 (n. 81). *See also* Nature; Wilderness
Williams, Agnes, 152
Williams, Raymond, 1, 21–22, 26, 45, 180 (n. 8)
Willow, Alaska, 47, 48
Wilson, Alexander, 5, 12, 97, 100

The Winds of Chance (Beach), 76–79
Wiseman, Alaska, 45
Wise-Use Movement, 138
Wister, Owen, 84, 190 (n. 66)
Winner, Langdon, 19
World in His Arms, The (Beach),
   75–76

Yosemite, 33–34, 41
Young, Samuel Hall, 30–34, 111
Yukon Territory, Canada, 61–62, 64,
   66, 67, 69, 70, 76, 78, 80
Yukon River, 64, 119, 149

Zhirinovsky, Vladimir, 89